Advancement of Phytoremediation Efficiency

Advancement of Phytoremediation Efficiency

Khushboo Chaudhary
and Dr Suphiya Khan

Delve Publishing
www.delvepublishing.com

Advancement of Phytoremediation Efficiency

Khushboo Chaudhary and Dr Suphiya Khan

Delve Publishing
224 Shoreacres Road
Burlington, ON L7L 2H2
Canada
www.delvepublishing.com
Email: orders@arclereducation.com

ISBN: 978-1-77407-807-5 (Hardcover)

ABOUT THE AUTHORS

Dr. Khushboo Chaudhary is presently working as a Technical officer-I in Translational Health Science and Technology Institute (THSTI) Haryana, India and having 1 year of teaching experience. Previously, she worked on "Improvement of Phytoremediation efficiency of Fluoride". She has published several research papers in international and national journals. She has published three international textbooks. She has got seven best paper and poster presentation award by Indian Society of Genetics and Biotechnology Research and Development and received the president appreciation awards also in the International conference. She has got the best poster award by ISSGPU Central Institute Research on Goats, Makhdoom. She has published several gene banks in NCBI Pubmed. She has also published a research article in a virology journal. She is likely to be a co-author in several of the publications and coauthor in J. Virological Methods.

Dr Suphiya Khan is working as a Associate Professor in the department of Bioscience and Biotechnology, Banasthali University, India. She has twelve years of teaching and research experience. Her strong background mainly relates to DNA fingerprinting, chemoprofiling, development of nano-adsorbents and Fluoride (F) phytoremediation technology. She has received various awards viz DBT-research Associateship, young scientist By ISGBRD, ICAR, recognition award for research and teaching and Indian National Academy of Sciences(INSA) international visiting scientist fellowship. She has recently selected as INSA visiting scientist for Turkey. She has been awarded as a principle investigator and co-investigator in various projects duly funded by UGC, MHRD, DST and DBT. Currently, she is handling Center of excellence on Water and Energy which is duly funded by MHRD with 2.5 crore rupees. Her work has been recognized internationally at various scientific conferences and journals. She has publishes over 30 peer reviewed articles and published two books on DNA fingerprinting and chemoprofiling. Dr Suphiya Khan's research interest focused on development of cost effective defluoridation technology for rural as well as urban people. She served as a reviewer of peer-reviewed journals in the area of DNA fingerprinting and Fluoride contamination. In addition to research, Dr Suphiya Khan is actively involved in teaching of undergraduate and post graduate students.

TABLE Of CONTENTS

LIST OF FIGURES

LIST OF TABLES

Table 4.22.2 Mineral content in soil of *P. juliflora* seedlings were grown in soil for 120 days in different concentration of F viz. control, 25, 50, 75 and 100 mg kg^{-1}NaF after treatment with 20 mM kg^{-1} Ethylene diamine tetraacetic acid in field experiment.

LIST OF PLATES

LIST OF ABBREVIATIONS

BF	Bioaccumulation factor
Ca	Calcium
CAT	Catalase
Chl a	Chlorophyll a
Chl b	Chlorophyll b
Cu	Copper
DW	Dry weight
EC	Electrical conductivity
EtBr	Ethidium bromide
F	Fluoride
Fe	Iron
FW	Fresh weight
g	Gram
hr	Hour
K	Potassium
KVK	Krishi vigyan kendra
LOD	Limit of detection
Mg	Magnesium
Mg/l	Milligram per liter
Mn	Manganese
N	Nitrogen
P	Phosphorus
PCR	Polymerase chain reaction
POD	Peroxidase
ppm	Parts per million
RDW	Root dry weight
RFW	Root fresh weight
SD	Standard deviation

SDW	Shoot dry weight
SFW	Shoot fresh weight
SPSS	Statistical product and service solution
SOD	Superoxidase
TF	Translocation factor
TISAB	Total ionic strength adjustment buffer
Total Chl	Total chlorophyll
WHO	World health organization
Zn	Zinc

PREFACE

The present book entitled "**Advancement of Phytoremediation Efficiency**" is aimed at presenting precise information for undergraduate, post graduate students and also very helpful to research scholars. This book has updated information's on phytoremediation management by using microbes and chelating agents, Fluoride contaminated soil affected crop productivity. It will be a milestone in the current scenario of environmental research on Fluoride pollution in groundwater, soil and its management. This book is an updated document on current research, the latest and cost -effective green technology like phytoremediation. Plants are valuable resources for all living organisms which provide food, medicine, produce oxygen and regulate the water cycle. Fluoride stress has a negative impact on the environment. Its has a direct and indirect effect on human beings and the drastic effect on crop yield. This book describes the development of cost-effective, sustainable and user-friendly technology for the farmers.

On the one hand, there is a vast scope for increasing plant nutrient supply through the use of microbes and chelating agents. All five chapters are an overview on Fluoride contaminated soil profile characteristics, macro and micronutrients in soils at Farmer's field, crop productivity and sodic soil quality for improving soil quality, use efficiency and crop productivity, plant- microbe interactions, phytoremediation approaches technique improving agriculture land, role of plant growth- promoting rhizosphere to improve soil fertility, fluoride effects on environment and using different amendments for improving crop productivity. Remediation is the only way to clean up the contaminants in soils and groundwater. Remediation refers to the process of environmental cleanup of contaminated sites and the techniques to reduce or eliminate contamination from soil or groundwater. General technical methodologies include the transfer of contaminants solely by themselves, with contaminated soils or groundwater to other places for final treatment or disposal; confinement of contaminants in place and destruction of contaminants in place. Remediation pathways include the transfer of contaminants alone or with contaminated groundwater to another place for final treatment or disposal, confinement, and destruction of contaminants in place. Soil remediation will be the primordial, burning and vexing issue of the present day and future of human civilization.

The preparation of this book, I have been greatly helped by several books and publications which I express my acknowledgment in the references and if I have omitted any references in my notice, please ignore these mistakes. This book is an updated document on current research and latest and cost-effective use of phytoremediation technologies. This book is a compilation of maximum information regarding the management of soil after using microbes and chelating agents.

1

Introduction

Fluoride (F) contamination is a worldwide problem. Many parts of the world are severely affected by this contamination and still, there is no cure of fluorosis (Kumari and Khan, 2017; Singh *et al.,* 2018). Fluoride is the main cause of environmental pollution in water, soil and vegetation. Fluoride form complex with some other elements in water and then converted into the salts and finally move in sediments of soils (Molla *et al*., 2007). Fluoride present at a low amount (1.5 mg L^{-1}) in drinking water can help in various physiological effects on human health and present at high concentration is known to cause dental and skeletal fluorosis (Teotia and Teotia, 1994). Fluoride at a high amount in the soil can cause various forms of toxicity which have negative effects on animals, plants and human health through the food chain (Stevens *et al*., 1997). Fluoride is naturally occurring elements consider in halogen family which is pale yellow gas extremely more reactive with other elements. It is always present in the compound form called Fluorides. Therefore it is well recognized widespread, non-biodegradable pollutants that accumulate in the soil, plants, animal tissue cell and probably most toxicity to a human being (Agalakova and Gusev, 2012).

Fluoride is absorbed by plant roots and then transported via the xylem to different parts of the plant (Pant *et al*., 2008), where it can get accumulated. The effect of fluoride on germination, physiological and biochemical parameters in different plant species have been studied by many researchers. These studies revealed that display to elevated fluoride can cause decreased germination, retarded plant growth (Miller *et al*., 1999; Sabal *et al*., 2006), chlorosis and leaf necrosis (Elloumi *et al*., 2005).

However, in fluorosis endemic areas, the fluoride content in plant parts have shown higher concentrations (Gupta *et al.*, 2009). The risk for human health and the environment can largely be affected by the concentration of fluoride that occurs in groundwater and the rate by which fluoride transfer to groundwater as both these processes can be strongly persuaded by the interaction of dissolved fluoride, with the soil solid phase via adsorption and desorption (Daniel *et al.*, 1985). Fluoride is entering the human body through drinking water, consumption of food and beverage (Malde *et al.*, 2011). It causes diseases such as hypertension and fluorosis due to the accumulation of vegetables (Koblar *et al.*, 2012). Ingestion of plants by animals, humans with a high F concentration can cause chronic toxicity, bone damage and tooth wear (Abugri and Pelig-Ba, 2011).

Researchers investigate that F concentration in drinking water present at a high amount has been found in Mexico, Holland, Itlay and Spain in South and North American countries (Mella *et al.*, 1994). India has been critically affected by high F concentration and 17 out of 29 states and Union territories were severely contaminated especially in Rajasthan (Vikas *et al.*, 2013). In India, about 20% of F concentration was found in the household water supply, out of this 10 % was only found in Rajasthan (Hussain *et al.*, 2010).

The conventional remediation technologies (reverse osmosis, forward osmosis, evaporation, ion exchange, adsorption and precipitation) are very expensive which do not remove F contaminants (Patel *et al.*, 2005). Therefore, there is a need to develop economically and effective methods to F decontaminate soils and minimize the F contamination for maintaining good health of the large population of the world. In recent years, using plants to remove contaminants from soil and water has gained significant importance. Phytoremediation is low cost -effective, most popular technology for the extraction, immobilization, stabilization, volatilization and degradation of contaminants (Abdul and Schroder, 2009; Al-Qurainy and Abdel-Megeed, 2009). It can be up to 1000-fold cheaper compared to conventional methods such as flotation-filtration, evaporation, ion-exchange, electrodialysis and ultrafiltration. It has been estimated that phytoextraction to clean up one acre of sandy loam soil to a depth of 55 cm will cost 60,000-100,000$ compare to 400,000$ for conventional using traditional soil removal methods (Salt *et al.*, 1995).

Microbe-assisted phytoremediation ,as well as rhizoremediation, appear to be most effective for the degradation of organic contaminants from impacted soil mainly when used in combination with appropriate agronomic technique (Zhuang et al., 2007). Plant growth- promoting rhizobacteria promote mechanisms such as the production of growth- promoting substances, siderophores and detoxification/removal of contaminates (Linger *et al.*, 2002). The mechanism of hyperaccumulation pant described briefly that hyperaccumulation does not only depend on the plant but also the interaction between microbes and the present amount available of metals and non metals in soil. Bacterial communities present in soil with tolerable metals like zinc, lead, copper and nickel. Previous study was reported that the rhizosphere of hyperaccumulating plants such as *Arabidopsis murale* and *Thlaspigoesingense* have an increased proportion of metal resistant bacteria. Many bacteria have a tendency that can alter heavy metal mobility for uptake plant easily (Rajkumar *et al.*, 2010). Soil bacteria produce such compounds biosurfactants, siderophores and organic acids which stimulate metal bioavailability in soil and increase root absorption of various ions like Mn^{2+}, Fe^{2+} and Cd^{2+} (Ali *et al.*, 2013; Emamverdian *et al.*, 2015). The selection of bacteria with the capability to tolerate and biodegradation of toxic compounds is very important. Pseudomonas *sp.* was able to tolerate heavy metal stress such as to degrade xylene, toluene and benzene at high concentrations. The previous study suggested that Pseudomonas *sp.* strains resulted in positive and useful for bioremediation purposes (Carla*et al.*, 2012). Therefore to our knowledge, no studies to date have evaluated in the case of F contaminated soil. However, selected Pseudomonas *sp.* was a good approach to the bioremediation of F contaminated soil for the present study.

Chelating compounds are a chemical substance composed of metal ions whose molecules can form several bonds with single metal ions. It is used as a micronutrient fertilizer and maintains its solubility with metal (Socha and Guerinot, 2014). Chelating assisted phytoremediation demonstrated as a possible treatment for remediation of heavy metal contaminated soil and sediment (Ullah *et al.*, 2014). The chelate compounds are being used to improve the quality of soil by increasing uptake and translocation of non-metals for the growth and development of the plant (Komarek *et al.*, 2007). Previous reports suggest that a great number of chelating agents have been used for enhanced phytoextraction (Ethylene diamine tetraacetic acid-EDTA, citric acid and malic acid). Ethylene diamine tetraacetic acid and citric acid as a chelator is considered as important in chelation therapy (Komarek *et al.*, 2007).

Ethylene diamine tetraacetic acid was reported as a chelating agent for the enhancement in the phytoextraction process (Blaylock *et al.*, 1997). It is well-organized and successful chelating agents for increasing the solubility of heavy metals in soils.Ethylene diamine tetraacetic acid has a different level of efficiency in soil and somewhere decreased in competitive cations for trace metals (Tandy *et al.*, 2004).

The direct uptake of metal-chelates and their translocation in shoots with the help of root surface and it was found that glycophosphate increases the Pb-accumulation in tested crops (Mathis and Kayser, 2001). The application of one large dose of chelates results to the extraction of more toxic metals (Finzgar and Lestan, 2007). Plants can remove Pb between 180 and 530 kg ha^{-1}per year, with the help of applying EDTA. According to Salido *et al.* (2003) 10 mM EDTA of 1kg soil was used to remove Pb from soil. *Brassica juncea* plants extract approximately 32 mg kg^{-1} of metal (Gleba *et al.*, 1999). Studies suggested that the application of microbe's to the plant rhizosphere can enhance the heavy metal uptake efficiency of the plant in field- scale experiments (Lominchar *et al.*, 2015).

It has been studied that *Prosopis juliflora*is a suitable species for phytoremediation of F polluted soils (Saini *et al.*, 2012). *P. juliflora* is a leguminous, perennial phreatophyte tree, distributed in arid and semi-arid regions. This is also naturally grown in F widespread areas of Rajasthan (India) without showing any morphological distortion and tolerant to very high temperatures (like 48°C). The tree can grow in a different area, as well as those with saline, alkaline, sandy, and rocky soils and the roots enter to large depths in the soil. *P. juliflora* is recognized to be tolerant of heavy metals and non-metals e.g cadmium, chromium, copper, fluoride (Kumar *et al.*, 2005). The number of nutrient elements in *P. juliflora* is the highest among the leguminous trees and the majority of accumulation power capacity for more biomass production by tree plants under responses to a high concentration of pollutants (Daldoum and Musa, 2012). The hyperaccumulator *P. juliflora* plant was taken for the present study. The present study intends to enhance the phytoremediation efficiency of *P. juliflora* for F contamination using microbes and chelates.

In the view of the above discussions, the following objectives are proposed for this purpose:

1. Isolation and characterization of Fluoride-tolerant bacteria from the rhizosphere region of *P. juliflora* and their potential in promoting the growth and F accumulation.
2. Effect of microbe inoculation on the activity of biochemical contents and efficiency of F uptake by *P. juliflora*.

3. Chelate assisted F uptake and its effects on biochemical parameters of *P. juliflora*.
4. Role of chelate assisted and microbe inoculation on mineral nutrition of *P. juliflora* and treated soil.
5. Remediation of Fluoride contaminant soil through microbe and chelate-assisted phytoremediation using *P. juliflora*: A field study.

2

Review of Literature

CONTENTS

2.1 FLUORIDE ORIGIN, HISTORY AND DISTRIBUTION

Fluoride(F) contamination is a worldwide problem. F is well recognized widespread, non biodegradable and relatively persistent pollutant element (Lee, 1983). F is a non-metal thus occurring naturally in air, soil and water as the F ion or bound in minerals or soils (Landis *et al.*, 2011). A high amount of F is present in igneous rocks generally, most of the F is found in sedimentary rocks (Weinstein and Davison, 2004).

As F is a natural component of soils, all organisms up to a certain extent exposed to F. Among the three forms of environmental media (air, soil and water), groundwater is a major source to exposure in humans (Landis *et al.*, 2011).

The World Health Organization and Indian Council of Medical Research described the drinking water quality guideline value for F is 1.5 mg/l (WHO, 1963; ICMR, 1975) Figure 2.1. Problems with high F concentration in the groundwater are mainly seen in Africa and Asia particularly in India.

High concentration of F in groundwater is often restricted to certain geological layers and the problem with contaminated drinking water can be very local, two villages situated near each other are not necessarily affected by the same natural pollution (Nielsen, 2009).

Fluorine is the 13th most common element in the crust of the planet approximately 0.032 % (Mason and Moore, 1982).

It is one of the 92nd naturally occurring elements. It is a member of the halogen family and pale yellow gas which extremely high electronegativity in the periodic table (Neumuller, 1981) and it is extremely reactive to another element. As a result, it is never found free in nature but only combined with other elements (Vikas *et al.*, 2009). Fluoride pollution spread all over the world, India is severely suffering from its effects (Meenakshi and Maeshwari, 2006).

According to a survey, high concentration of F in drinking water has been found in South and North American countries, Itlay, Holland, Mexico, Algeria, Brazil, Canada, China, and Spain Figure 2.1 (Mella *et al.*, 1994; Mirlean and Roisenberg, 2007; Messaitfa, 2008; Desbarats, 2009). In India, 17 out of 29 states and Union territories are critically affected by high F concentration (Vikas *et al.*, 2013).

Rajasthan having arid to the semi-arid environment, is the severely F affected state. In India, about 20% of F concentration was found in the household water supply, out of this 10 % was only found in Rajasthan

(Hussain *et al*., 2010). Newai Tehsil (Tonk district), located in the southeast region (N 26°23', E 73°8') of Rajasthan, India, has a semi -arid climate and various kinds of stress conditions viz. low and erratic rainfall, drought, and F (Govt. of Rajasthan,2005; 2010). Study groundwater quality and pollution problems in groundwater of the Newai Tehsil area have also been studied (Baunthiyal, 2008). Most of the people in this area suffer from dental and skeletal fluorosis such as mottling of teeth, deformation of ligaments, bending of the spinal column, and aging problem (Yadav and Khan, 2010).

Some foodstuffs such as vegetables and fruits normally contain F though at low concentration (0.1 mg kg^{-1} to 0.4 mg kg^{-1}) and thus contribute to F intake by man.

Higher levels (up to 2 mg kg^{-1} of F) have been found in barley and rice (Chakarbarti *et al*., 2013).

Fluorine has the ability to make strong hydrogen bonds. Of all metal ions, Al^{3+} makes the strongest bonds to F^- but also beryllium binds with high affinity (Li, 2003). Fluoride is mainly bound in complexes with either aluminium or iron (e.g. AlF_2^+, AlF_2^+, AlF_3, AlF_4^-, FeF_2^+, FeF_2^+, FeF_3) (Elrashidi and Lindsay, 1986).

The highest levels of fluoride are associated with felsic and biotite gneisses, syenites, granodiorites, granites, alkaline volcanic and quartz monzonites (Chae *et al*., 2007).

According to the South Carolina Ambient Groundwater Quality Report (2003) have found F concentration in water is greater than 3.5 mg L^{-1}, where 40% of people are depending on groundwater for their needs.

The high concentration of F in groundwater up to 6.27 mg L^{-1} has a better correlation with well depth of volcanic ash deposits in Texas (Hudak and Sanmanee, 2003).

The sources of F in sediments and sedimentary rocks include argillaceous deposits, phosphate beds, carbonate rocks, marine sediments, shales, parent rocks, F rich clays and volcanic ash (Chae *et al*., 2007).

Most of the world's sporadic incidence of high F contents present in drinking water has been reported from South and North American countries, Mexico, Itlay, Spain, Holland, Pakistan, China;

South Korea, Sri Lanka, and India (Guo *et al*., 2007; Kim *et al*., 2011; Vikas *et al*., 2013). Fluoride is found as 85 million of tons on the earth's crust and deposits, out of its 12 million in India (Teotia and Teotia, 1994).

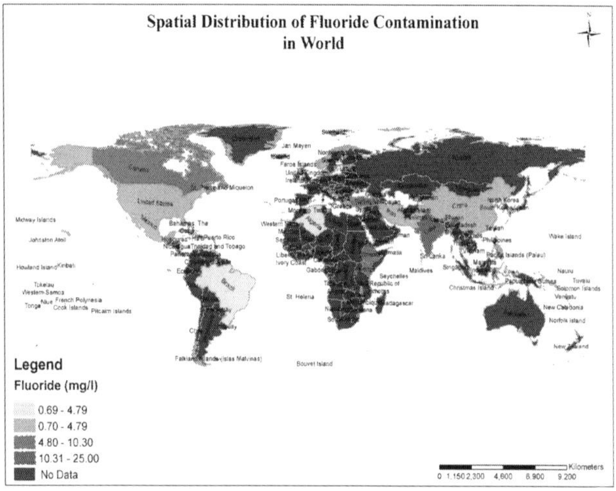

Figure 2.1: Worldwide distribution of F contaminants in groundwater (Mella *et al.*, 1994; Guo *et al.*, 2007; Mirlean and Roisenberg, 2007; Messaitfa, 2008; Desbarats, 2009; Kim *et al.*, 2011; Vikas *et al.*, 2013).

Naturally, F-contamination is spread in India and causes an excessive problem with human health. In India, F is present at a high amount in groundwater was reported first time in 1937 in the state of Andhra Pradesh (Shortt *et al.*, 1937). The total population has consuming drinking water with a high amount of F which is over 66 million peoples according to (FRRDF, 1999). Some regions in northwestern and south India are affected by fluorosis (Yadav *et al.*, 1999). The groundwater exceeds its maximum limit of F in drinking water in Delhi and its surrounding area(Datta *et al.*, 1996; Joshi, 2003; Nandi, 2013). The rocks in South India are rich with F which forms the main reason for contamination of groundwater especially in the Nalgonda district (Andhra Pradesh). According to Mondal *et al.* (2009) rocks are enriched in F from 460 to 1706 mg kg^{-1} in Kurmapalli watershed.

Different states in India have been reported for a number of cases about fluorosis from the granite and gneissic complex such as Tamil Nadu, Delhi, Orissa, Karnataka, Rajasthan, Kerala, Haryana, Bihar, Gujarat, Andra Pradesh, Madhya Pradesh, Maharashtra, Jharkhand; Guwahati, Uttar Pradesh and West Bengal Figure 2.2(Singh and Singh, 2000; Das *et al.*, 2003; Khandare, 2006; Sreedevi *et al.*, 2006; Shaji *et al.*, 2007; Kundu and Mandal, 2009; Brindha *et al.*, 2011; Kotecha *et al.*, 2012; Pandey *et al.*, 2012; Vikas *et al.*, 2013). Rajasthan is the largest state of country having 10.14% of the country's area, 5.5% of the nation's population but has low water resources and an area of 3.4 lakh sq. km cover in India. Most of the parts of the state have excess F in saline soil. Rajasthan state is endemic for F and high F concentration in groundwater is influenced by the regional and local geological setting and hydrological conditions for F contamination (Agarwal *et al*., 1997). The Sanganer area of Jaipur District groundwater quality was found that 75% of the villagers are suffering from dental and skeletal fluorosis (Gangal, 2007). Health- related issue in 23 districts of Rajasthan has become a serious problem among all 31 districts (Datta *et al*., 1999). Tonk district (Newai Tehsil) located in the southeast region (N 26°23', E 73°8') of Rajasthan, India. It has semi-arid climate and various kinds of stress conditions viz. low and erratic rainfall, drought and F (Govt. of Rajasthan, 2005; 2010). The groundwater quality and problems in groundwater of the Newai Tehsil area have been studied in the Banasthali Vidyapith laboratory (Baunthiyal, 2008). Most of the people in this area suffer from dental, skeletal fluorosis (such as mottling of teeth, bending of the spinal column, deformation of ligaments and aging problem (Yadav and Khan, 2010).

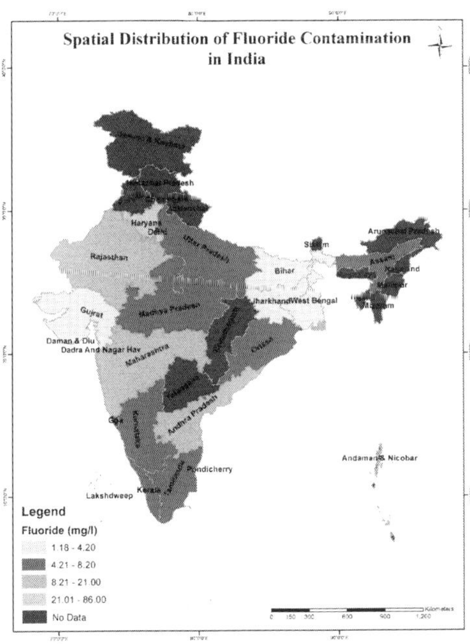

Figure 2.2: Statewise distributions of F contaminants in groundwater (Singh and Singh, 2000; Das *et al*., 2003; Khandare, 2006; Sreedevi *et al*., 2006; Shaji *et al*., 2007; Kundu and Mandal, 2009; Brindha *et al*., 2011; Kotecha *et al*., 2012; Pandey *et al*., 2012; Vikas *et al*., 2013).

2.2 SOURCES OF FLUORIDE (F)

Fluoride is released into the environment through wastewater, coal combustion and waste from various industrial processes such as adhesive production, phosphate ore processing, copper and nickel production, glass brick and ceramic manufacturing etc. Figure 2.3 (Stevens *et al.*, 1997). F is native to the soil within several minerals most commonly apatite especially forms of topaz [$Al_2(SiO_4)F_2$], fluorite (CaF_2), cryolite [(Na_3AlF_6)], fluorapatite [$Ca_5(PO_4)_3F$] and micaceous clay minerals. The total F concentration in normal soils ranges from 150 to 360 ppm but can reach up to 620 ppm. The increasing concentration of F uptake is toxic to plants and animals so it is desirable to reduce the rates of F accumulation in soils (Madhvan, 2001; Jha *et al.*, 2011). The maximum content of F in groundwater is usually controlled by the solubility of fluorite (CaF_2). The natural sources of F such as weathering of rocks and minerals together with releases from volcanic activity according to WHO (2002). Madhavan and Subramanian (2001) suggested that certain rocks ranging from greywaeke to 800 $\mu g\ g^{-1}$ in granite and 180 $\mu g\ g^{-1}$ in sandstone. Most developing countries are strongly affected by the application of fertilizers and industrial airborne pollutants (Brindha and Elango, 2011). The airborne F in the vicinity of aluminum factory, phosphate fertilizer factory and glass fiber factory has been often contaminated edible crops and dietary vegetables along with surface water bodies (Pandey, 2005).

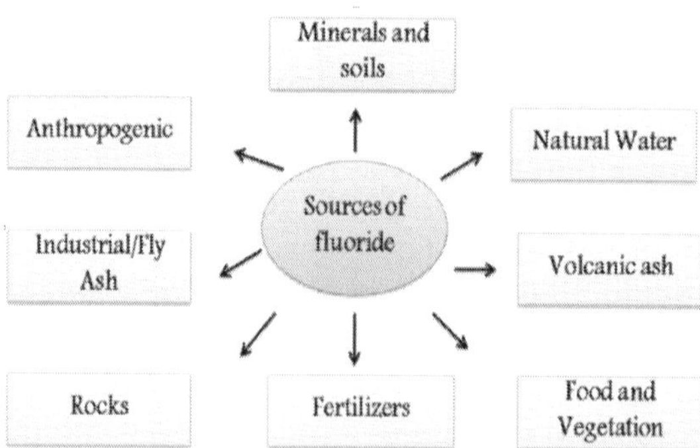

Figure 2.3: Sources of Fluoride.

2.3 EFFECT OF F ON PIGMENTS AND ENZYMES

Fluoride adversely affects the growth and productivity of crops. The reduction of root and shoot lengths was due to the unbalanced nutrient level of plants in the presence of F (Datta *et al.*, 2012). The effect of F induces morphological symptoms like chlorosis and necrosis, which soon changes in color turning yellow reddish and end in necrosis in leaves (Landis *et al.*, 2011). The chlorophyll content was reduced leads to yellowing of the attacked plant tissue and necrosis occurs when plant tissue starts to die (Landis *et al.*, 2011). However, fluorine is the most electronegative element which can also bind with Mg^{++} ion and forming complexes reported as metal-fluorine compounds. So, it is effect on plant metabolism, biosynthetic turnover of photosynthetic pigments at low concentration of F.On the other hand, to resist this damage plants have their self antioxidative defense system (Chugh *et al.*, 2011; Bauddh and Singh, 2012). Fluoride has prolonged metabolic inhibitor interfering with the metabolism of carbohydrates, lipids and proteins (Reddy and Venugopal, 1990). The most important antioxidant enzymes are catalase (CAT), peroxidase (POD) and superoxidase (SOD) (Changcheng *et al.*, 2012). The chlorophyll content decreases with an increase in F concentration due to inhibition by F of incorporation of aminolevulinic acid into chlorophyll synthetic pathway Figure 2.4 (Bhargava and Bhardwaj, 2010; Saini *et al.*, 2012). Therefore, there is a need to develop an economically and more effective F remediation method.

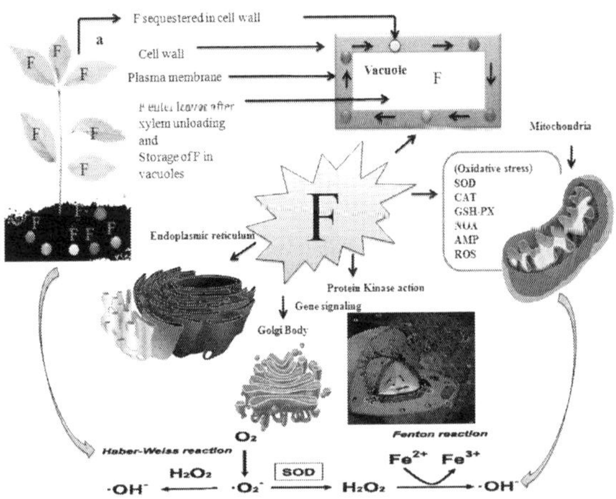

Figure 2.4: Fluoride toxicity mechanism in a plant.

The available methods for F removal are reverse osmosis (RO), forward osmosis (FO), evaporation, ion-exchange, adsorption and precipitation (Sehn, 2008; Yao *et al.*, 2012; Samadi *et al.*, 2014). Precipitation involves the precipitation of sparingly soluble F salt as insoluble fluorapatite by the addition of chemicals (coagulants and coagulant aids). The most common materials used in the technique are aluminum salts (e.g. Alum), lime and poly aluminium hydroxy sulfatesulfate. The best example of precipitation technique is Nalgonda technique developed by the National Environmental Engineering Research Institute (NEERI), Nagpur (Ashraf *et al.*, 2016). Adsorption is mainly based on the adsorption of F ions on the surface of an active agent. The commonly used adsorbent materials for this process are activated alumina, rice husk, bone charcoal, activated carbon and serpentine (Sabir *et al.*, 2015) Toxic waste produced, highly pH dependent, adsorbents develop fouling smell, effectiveness decreases with time. The common membrane processes for F removal are reverse osmosis (RO), nano-filtration and electrodialysis. Fluoride ion is not removed due to its small size (Sehn, 2008). Ion-exchange processed synthetic chemicals, namely, cation and anion exchange resins (Samadi *et al.*, 2014; Khalid *et al.,* 2017).Fluoride removal techniques due to its high cost and complexity are still not satisfactory. The conventional technologies are too expensive and costly so it is important to develop an eco-friendly more effective method to decontaminated soils. Many conventional methods were used to remove contaminants but not successful due to the high cost and least effectiveness. Recently using plants for F contamination has gained importance. Phytoremediation technology uses different kinds of plants for the removal of contaminants from the soil and water, it provides ecologically and environmentally safe method for remediation (Chaudhary *et al.*, 2016; Kumari and Khan, 2017). The cost for conventional actions estimated such as vitrification and soil washing, as between US$ 100000 and 10,00000 per ha (Russel *et al.*, 1991), whereas the cost of phytoremediation was judged between US$ 60000 and 100000 per ha (Salt *et al.*, 1995).

2.4 PHYTOREMEDIATION TECHNOLOGY

Phytoremediation is the use of plants for environmental remediation and involves removing organic compounds and metals from soils and water. This technology is based on plants that have the ability to tolerate high levels of heavy metals (Salt *et al.*, 1995). Phytoremediation involves a number of biological mechanisms including direct uptake, the release of exudates into the rhizosphere (to enhance bacterial and fungal processes), and metabolic processes within the root and shoots cells (Singh *et al.*, 2003). Selected plants are grown on the contaminated site, where they draw up pollutants and concentrate them within various tissues. The plants are then harvested and may be further treated by burning them in a controlled system. The residue of the plants could be recycled or placed in landfills (Ma *et al.*, 2011). Phytoremediation involves a number of biological mechanisms including direct uptake, the release of exudates into the rhizosphere (to enhance bacterial and fungal processes), and metabolic processes within the root and shoot cells.

Selected plants are grown on the contaminated site, where they draw up pollutants and concentrate them within various tissues (Rajkumar *et al.*, 2010).

Designing a phytoremediation system varies according to contaminant conditions at the site, the level of clean up required and the plant species to be used (Ali *et al.*, 2013).

It is a broad term that involves several different techniques such as, Phytofiltration, phytovolatilization, phytodegradation, phytostabilization and phytoextraction Figure 2.5(Ali *et al.*, 2013). Phytoextraction is a very useful process in which contaminants in soils are uptake through the roots and translocation into the aerial parts of the plant. Nature has not a tendency that all plant accumulates in equal proportion. Some plants have accumulated in large quantity and these hyperaccumulator plants are the base of phytoremediation technology.

There are different steps that involve in phytoextraction process (i) uptake and bioavailibity (ii) translocate of heavy metals (iii) sequestration of metals in leaves and vacuoles. A high amount of heavy metals concentration accumulates in plant organs is not usually a natural process for favored reaction somehow it's the plant capability to uptake more than other plants (Greipsson, 2011)(Figure 2.5).

Plant defense system mechanisms play a role in metabolic, physiological and expressional changes under stressful conditions caused by different pollutants. Germplasm of hyperaccumulators is the backbone of this technology.

Therefore, understanding the genetics of hyperaccumulation is an important tool for the enhancement of hyperaccumulation efficiency. Phytochelatins (PC) and metallothionines (MT) and heavy metal ATPase (HMA) genes play a crucial role in signaling, uptake, detoxification and accumulation of metal. Their combined role enhances the hyperaccumulation efficiency. This technology helps for the development of plants with the higher potential to clean our environment by giving a favorable condition (Chaudhary and Khan, 2018; Chaudhary *et al.*, 2018).

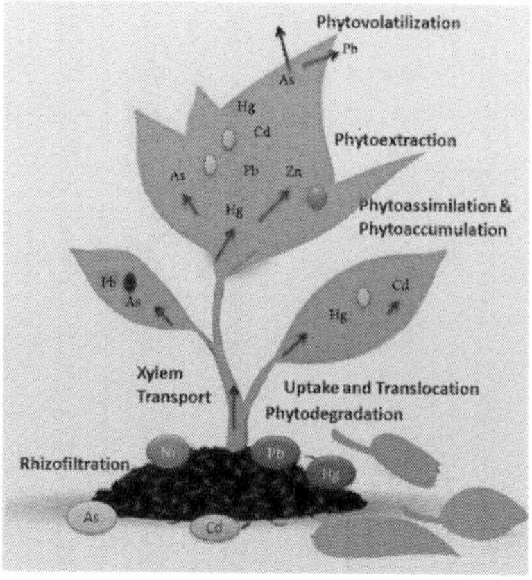

Figure 2.5: Types of phytoremdiation process.

Phytochelatins (PCs) are the most important class of metal chelators, it is used to chelate the variety of toxic metals. Phytochelatins are produced in a cell under the stress condition and reaction to the high concentration of heavy metals. Some important PCs-metal complexes have been derivative from different microorganisms, fungi, and plants. The work of PCs-metal complexes is to lower down the binding capacity of heavy metals to the cell wall and the same time to detoxify the cell compartments. It can resist the very high concentration of heavy metals without causing toxicity (Sunitha *et al.*, 2013; Sharma *et al.*, 2016; Chaudhary *et al.*, 2018). In assessment of free metal ions, the PCs-metal is in complex form and much more stable (Figure 2.6).

Metallothioneins (MT) are also a group of phytochelatins that binds heavy metals through a thiol group of cysteine and also plays an important role in detoxification of heavy metals (Jia *et al.*, 2012). These MT have a different mechanism to protect the plant from heavy metals by scavenging the ROS and sequestration (Huang and Wang, 2010). It regulates the action of metallodrugs, their transcription genes activation and the activity of metalloenzymes under any stress condition (Bractic *et al.*, 2009; Gautam *et al.*, 2012)). The regulation of MT genes was depending on the type of plant tissue (Yuan *et al.*, 2008). These genes are activated when plants under abiotic stress such as cold, heat, salt, drought, heavy metal and oxidative stress (Usha *et al.*, 2007; Usha *et al.*, 2009; Singh *et al.*, 2011). Metallothionein genes helped to keep the plant from metal by their hyperaccumulation and they are expressed in high concentration in hyperaccumulator plants as compared to non-hyperaccumulator plants (Gautam *et al.*, 2012).

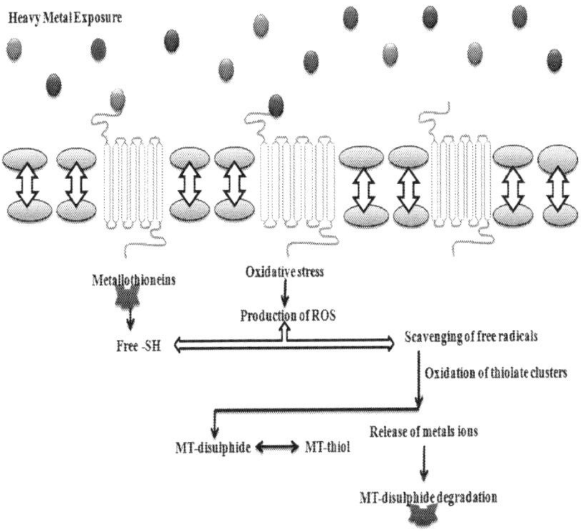

Figure : 2.6: Phytochelatins (*PCs*) genes expressed in various plants under heavy metal stress.

Themetallothioneins (*MTs*) are classified on the basis of cysteine arrangement in their structure. There are more than 20 conserved Cys are found in mammals and vertebrates which are known to their tolerance towards Cd ions (Shri *et al.*, 2014; Gu *et al.*, 2015).

Metallothioneins are classified into four classes first the *MT1*, it was expressed in a plant named *Cicer arietinum* and their subclasses *MT1a* and *MT1c* in *A. thaliana*.

The second one is *MT2*, it is also found in plant *Cicer arietinum* but their subclasses are *MT2a* and *MT2b*. And the last one is class *MT3*, it is found in *A. thaliana* and *Musa.* Some other classes include *MT4a-Ec-2* and *MT4b-as Ec-1* found in *A. thaliana* and *Triticum aestivum*(Hassinen *et al.*, 2011; Lee *et al.*, 2014).

There is strong structural similarity between GSH and *MT3* due to their same biosynthesis precursor molecule which is thiol-rich tripeptides.

A powerful inhibitor named buthionine sulfoxamine, inhibits the activity of g-glutamylcysteine synthetase enzyme which leads to the decrease in the concentration of g-glutamylcysteine and GSH in cells (Figure 2.7).

In *Silene cucubalus*, *MT3* was found which is synthesized formed by the reaction of GSH and g-glutamylcysteine dipeptidyl transpeptidase (Grill *et al.*, 1989).

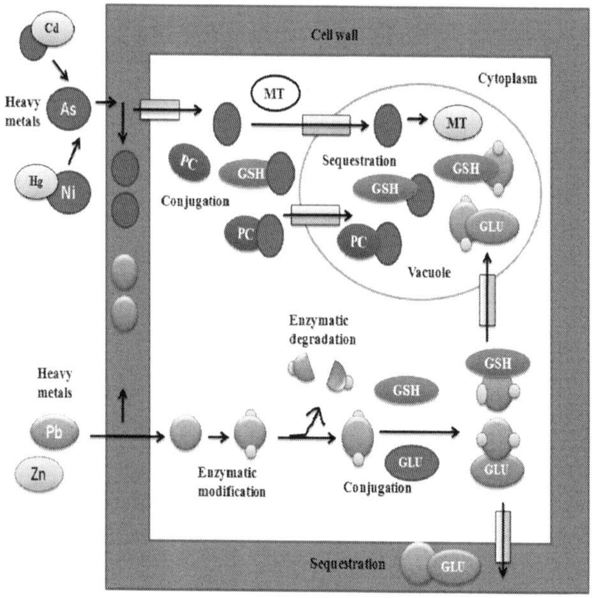

Figure 2.7: Diagram of detoxification, conjugation, and sequestration in the vacuole where the pollutant can do harm to the cell. (Chelators shown are GSH: glutathione; GLU: glutamate; MT: metallothioneins; and PCs: phytochelatins).

Significantly progress has been made in their recognition and their role in phytoremediation. Some of the other genes like heavy metal ATPases are also used for phytoremediation, these are *HMA2, HMA3, HMA4* (Chaudhary *et al.*, 2016). The expression of these genes is liable for heavy metal uptake, translocation, and sequestration might be allowing the yield of plants that can be effectively exploited in phytoremediation. The most closely connected to the *A. thaliana*isP1B-type genes. The *HMA2* is expressed in the translocation of Zn and Cd in *A. thaliana*, barley, rice, and wheat. In *Arabidopsis*, the cellular and subcellular patterns of *AtHMA2* expression were related to the*AtHMA4* gene.

The expression of *HMA2p-GUS* gene was observed for the most part in the vascular tissues of the leaf, stem, and root. *HMA2*-GFP proteins were also localized in the plasma membrane of the plant cell. Recently, results on the characterization of the *HMAs*2 gene from the different plants for possible application apply in phytoremediation approaches (Chaudhary *et al.*, 2016). The ATPase families of integral membrane transporter proteins that help to uptake transition metals are involved in mediating metal-resistant and metal-hyperaccumulating traits. The plants were expressing 35S promoter

*AtHMA4*as well also the metal transporters such as *HvHMA2*(Barabasz *et al*., 2013). While *HMA3* may involve metal detoxification by sequestering Cd into the vacuole *HMA4* acts as a physiological master switch during the process of hyperaccumulation metal, and *HMA2 and HMA4* play roles in root to shoot metal translocation (Figure 2.8). It is hypothesized that the roles of metal transporters in plants will be essential for the development to genetically modify plants that accumulate specific metals, with subsequent use in phytoremediation process. The efficiency enhanced of *HMA3 and HMA4*is a prerequisite for hyperaccumulation and hyper-resistance in hyperaccumulators plant (Figure 2.8).

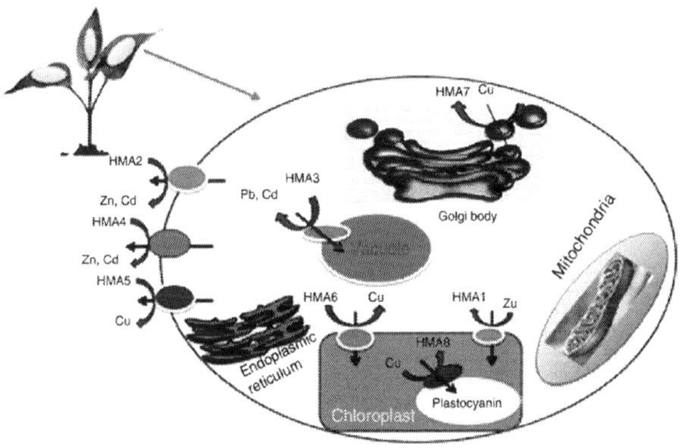

Figure 2.8: Heavy metal ATPase (HMA) gene contributes to hyperaccumulation of heavy metals.

2.5 HYPERACCUMULATORPLANTS

Hyperaccumulator term was proposed first time by Brooks *et al*. (1977) in reference to those plants that can accumulate more than their natural favored condition approximately 1000 mg kg^{-1} of heavy metals. Plants accumulate more and more contaminants and tolerate without showing any symptoms (Memon and Schroder, 2009). Baker and Brooks suggested that the minimum threshold tissue concentration for plants as 0.1% and considered Ni, Cr, Cu, Co, and Pb hyperaccumulators but same as above the experiment was done in case of Mn, and Zn threshold value for plants was established as 1% (Baker *et al*., 2000). Plants accumulate heavy metals in root to shoots which favored that can allow translocation of minerals and sugars as they require a proper ratio maintained between the amounts of heavy metals specific in roots to shoots. This process is named a translocation factor (TF).

Hyperaccumulator plants need more than the TF value 1 (Tangahu *et al.*, 2011). Same as above another factor discussed here bioaccumulation factor (BF) value also is required more than 1 for hyperaccumulation (Ahmadpour *et al.*, 2014).Some different 450-500 plants have been identified as hyperaccumulators which include *Thlaspi caerulescens* that accumulate (Pb, Ni, Cd and Zn), *Arabidopsis halleri* that can accumulate high levels of heavy metals (Cd and Zn but not Pb), *Alyssum bertolonii* can uptake (Ni and Co) and some other plants which belong to different families can also participate to accumulate heavy metals such as *Caryophyllaceae, Fabaceae, Poaceae, Lamiaceae, Asteraceae, Cunoniaceae and Cyperaceae* and many others (Maestri *et al.*, 2010). Plants have specific properties that give us some specific advantages to remediate the environment (Meagher *et al.*, 2000). Plants absorb metal particles through roots and root hairs that generate surface area through which pollutants can be extracted from contaminated soil and water. The plant is autotrophs; they take up nutrients directly from the environment in the gaseous form with the help of the photosynthesis process.

Heavy metals translocate in roots to shoots and also leaves. It depends on plants which are used for phytoremediation purpose and called as hyperaccumulators (absorb more than required) and non-hyperaccumulators when they do not absorb limited amount. Drawbacks are also considered when there is a significant reduction in the biomass of the plants. Different species are used to remove contaminants from the soil and water but sometimes there is an inability of plants mechanism to absorb insoluble form of heavy metals present in the soil. This process is dependent on many circumstances like soil pH, water contents and also the presence of organic and inorganic substances. Naturally, plants have the ability to uptake contaminants due to existing in soluble form in soil and water. However, other types of reaction can also be taken up by the use of different amendments like plant growth-promoting bacteria and also chelant-induce hyperaccumulation mechanisms around the root (Abollino *et al.*, 2006).

2.6 FLUORIDE ACCUMULATION PATHWAY IN PLANT

Fluoride accumulation in plant tissues followed by root>shoot>fruit and leaves(Fung *et al.*, 1999).F accumulation in the roots is higher than in leaves, stems and seeds (Weinstein and Davison, 2004). Most of the F accumulates in roots and then transports across the roots remain in the cell walls and intercellular spaces (apoplasts), rather than through the cell membranes and the endodermis (symplast) (Takmaz-Nisancioglu and Davison, 1988). The impermeable Casparian strips in the wall of the endodermis act as a barrier for F to enter the systems, which limit transport to the shoot and leaves. However, the casparian strip is discontinuous at the root tips and at sites of developing lateral roots. At these points, molecules can be cross into the xylem and can be conducted within the system. Different plants have tolerated power towards contaminants but it depends on calcium (present in the cell wall). Calcium acts as a buffer against F accumulation. Still, there is a lack of knowledge about the F toxicity pathway and mechanism.

2.7 PHYTOREMEDIATION OF F CONTAMINATED SOIL

Phytoremediation is the use of green plants to remove contaminates pollutants from soil, water, and air (Greipsson *et al.*, 2011). A search for F hyperaccumulators is a necessary process for phytoremediation in F-endemic areas. Four important characters used in defining a plant as a hyperaccumulator are (i) translocation factor (ii) bioconcentration factor (iii) tolerance(iv) enrichment factor (Lorestani *et al.*, 2011). It was found that the maximal F accumulation took place in roots (16.64-106.2 mg kg^{-1}) whereas in the edible part (fruit), it varied from 39.3 to 48.51mg kg^{-1} in the treatment range of 0-600 mg kg^{-1}NaF soil. Recently investigate the potential of eight tree species of semi-arid region for a hyperaccumulation of F (Baunthiyal and Sharma, 2008). Their result suggested the potential use of *P. juliflora* in F removal from groundwater and soil. Plants tolerant and resistant to F are a good candidate for remediating F from water and soil (Saini *et al.*, 2012).

2.8 FACTOR AFFECTING PHYTOREMEDIATION EFFICIENCY

2.8.1 Plant Growth- promoting Rhizobacteria

Bacteria have the ability to the tolerance of metals using a different mechanism which maintains metal homeostasis with keeping the concentration of essential metals (Nies, 2003). It involves both actively and passively for metal uptake, remaining or sequestering.The soil has numerous microbial inhabitants including fungi, protozoa, and algae, with the most pronounced being bacteria (Curl and Truelove, 1986). A high amount of heavy metals decrease the microbial activity in the soil and affect crop production by getting accumulated in plant organs. Metal ions present in the soil are first absorbed by roots and then transported to different organs. The plant proteins and enzymes present in cells have an affinity for heavy metals which render them inactive and make them lose their function. Once they interact with heavy metals, a change in protein structure occurs, which ultimately affects the plant growth by inactivation of the photosystem (Usha *et al*., 2009). There is a huge amount of proteins that can bind with different heavy metals with greater affinity. These metal -binding proteins are at the outer membranes in the plants and microorganisms where they interact with the metal ions present in the environment, thus ensuring the transport of these metals to the cytosol, which is then transferred to suitable receptor protein by the metallochaperones (Singh *et al*., 2011).

Bacteria play role to stimulate root growth, increased soil nutrients for obtainable to the plant, fix nitrogen, improve soil fertility, suppress soil-borne pathogens, hormone production, nodule formation, nutrients uptake, siderophore production N-fixation and protect from different diseases to important role in crop yield under various environmental stresses like heavy metals, drought, salinity and temperature (Naik *et al*., 2012). Bacterial species *Arthrobacter spp*. and *Streptomyces spp*. were dominant and could tolerate as much as 10 mM Ni in plate assays. *Pseudomonas aeruginosa* strain WI-1 could rate up to 0.6 mM lead nitrate and also accumulated lead at 26.5 mg g^{-1} of dry cell biomass (Naik *et al*., 2012). Bacteria that provide benefits to plants may be symbiotic bacteria, which are closely associated with plants (for example *Rhizobia*), and also bacteria that are free-living in the soil (Figure 2.9). Another factor for plant growth is ACC deaminase has been studied with respect to heavy metals stress, produced ethylene as an intermediate in the same favored reaction. Ali *et al*. (2013) reported that ACC deaminase can produce by bacteria at lower levels of ethylene in plants for promoting plant growth. Both plants and bacteria produce the auxin IAA, which increasing plant biomass and metal accumulation (Sing *et al*., 2011). Cr-resistant bacteria were enhancing plant growth by re-inoculation and less accumulation of Cr compare to un-inoculated sunflower plants (Faisal and Hasnain, 2005).

Many PGPB strains can synthesize a phytohormone, indoleacetic acid (IAA) that acts to enhance various stages of plant growth (Glick *et al*., 1998).IAA is taken up by the plant,where it can stimulate plant cell proliferation and cell elongation. It can also stimulate the activity of the enzyme 1-aminocyclopropane-1-carboxylic acid (ACC)synthase, which irresponsible for the conversion of S-adenosylmethionine (SAM) to ACC, a precursor for the plant stress hormone ethylene (Glick *et al.*, 1998). Bacterial strains that possess ACC deaminase can use as a source of nitrogen, effectively lowering the outside concentration of ACC(Glick, 2003).

Mechanism of phytoremediation technology and metal accumulation somehow depends on the chelating molecules synthesized by bacteria. The combination of bacteria interaction with some chelating compounds in specific plants may give more benefits to the plants for the better accumulation of contaminants from the soil with different processes such as bioaccumulation, bioleaching, biotransformation, biodegradation and biosorption. There are different processes such as bioleaching involving bacteria*Thiobacillusspp.* and *Aspergillus niger*, Fungus, biosorption of low concentrations of metals in water by algal or bacterial cells, bio-oxidation or bioreduction of metal accumulation by *Bacillus subtilisBerknolderia sp., Methylobacterium, Bacillus megaterium, Pseudomonas sp., Kluyvera ciscorbata, Brevibaccillus, Pseudomonas fluoreseus, Flavobacterium, Bacillus cereus, Bacillus sp., Berk molderica sp.*sulfate-reducing bacteria and biomethylation of heavy metals such as Ni, Pb, Zn, Cr, Hg, Cd, Cu. Heavy metals in soil that adversely affect the microbial population and also affected soil properties which leading to loss yield of crops (Emamverdian *et al.*, 2015). Most of the heavy metals are not degraded properly due to less mobility, hence persist in the environment. Bacteria have the ability to the tolerance of metals using a different mechanism which maintains metal homeostasis with keeping the concentration of essential metals(Nies, 2003). Microorganisms involve both actively and passively for metal uptake, remaining or sequestering. The effects of different bacteria on metal uptake, depend on the basis of chromosomally or extrachromosomally which controlled detoxification of metals (Felestrino *et al.*, 2017). Several type of sequestration mechanisms of metal resistant systems including efflux pumps to remove metals from the cell and to bind inside the cell. Metals pump out by using adenosine triphosphates two efflux system and through antiports proton generate protein gradient across the cell membrane (Nies, 2003). Researchers indicated another mechanism about metals resistance in cyanobacteria is sequestration by metallothioneins signaling. Metallothionein compounds bind to different metals on the receptor of the plasma membrane to sulfhydryl group of cysteine residue with phosphate groups and activate the channels for the expression of genes to tolerate under stress (Jia *et al.*, 2012).

Pseudomonas fluorescens strain has been documented and little information about the role of bioremediation. The strain was given positive results in heavy metals detoxification and also multi-metal stress. *P.fluorescens* appeared to favor the detoxification process and useful for bioremediation purposes (Azhar *et al.*, 2015).*P.aeruginosa* was found to be tolerant to different heavy metals concentration. It was able to grow under a wide range of pH and temperature. The strain was promoted for plant growth; increase the nutrients level in the soil. It has a high degree of heavy metals resistance potential such as in the case of As.*P. aeruginosa* was considered as bioremediation agents in heavy metals and F contaminated soil (Jariyal *et al.*, 2015; Chaudhary and Khan, 2016; Biswas *et al.*, 2017).

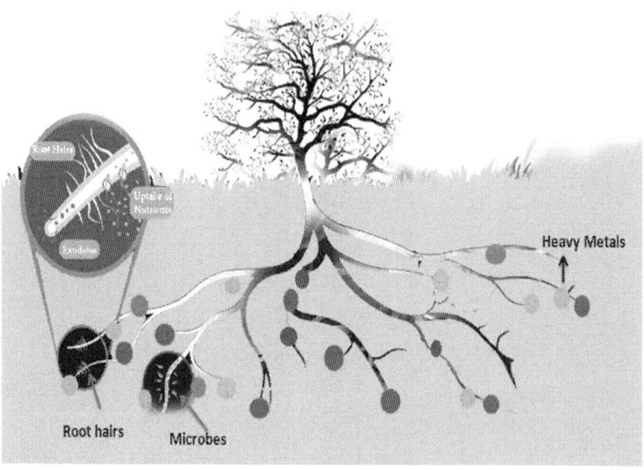

Figure 2.9: Plant-microbe interaction.

Inoculation of *P. aeruginosa* was found to enhance antioxidative enzyme such as peroxidase (180%), catalase (83%), and superoxidase dismutase (161%) and also the non-enzymatic components such as total phenolics (10%) and ascorbic acid (65%) compared with treated plants under Zn stress. Plants' roots produced some enzymatic activities to easily absorb contaminants of metals and mobilize metals ions for the bioavailable fraction so plants cannot easily absorb metals without the help of enzymes produce around the roots (Li *et al.*, 2007). *P. koreensis* AGB-1 strain increased SOD and CAT activities by 33% and 42% respectively. *P. aeruginosa* strains were isolated from contaminated saline soil and adversely effects on biodegradation of petroleum hydrocarbon as compared to the untreated soil. The strain was an effective means of reducing the salinity of contaminated soil and to be considered as microbial remediation. *P. aeruginosa* strain DN1 was isolated from petroleum-contaminated soil and showed excellent emulsification at 100%. The strain DN1was exhibited an increase in the capacity to degrade crude oil by approximately 90.52%. Previous results were indicated that the use of strain means very effective for bioremediation of crude oil (Kuang *et al.*, 2016). *P. putida* mt-2 strain was previously reported as effective for biomass growth and bioremediation strategies (Wang *et al.*, 2011).*P.putida* and *P. stutzeri* was isolated from the phenol and cyanide rich contaminated soil. These two isolates strains were tolerated up to 1800 mg L^{-1} phenol and up to 340 mg L^{-1} cyanide concentrations.

There are different mechanisms such as complexolysis, redoxolysis and acidolysis which related to the capacity of microbes to oxidize the structural elements of minerals, to decrease soil pH and to increase the concentration of metal chelators (Brandl and Faramarzi, 2006). In complexolysis process, complexes are formed at the surface of metal-bearing phases between metals and microbial chelators (Grybos *et al.*, 2011). Complexolysis process is also known as ligand-induced metal solubilization.Redoxolysis involves bacteria that obtain energy from solid-phase minerals (Figure 2.10). Throughout acidolysis, also called proton-induced metal solubilization, protons secreted by microbes bind to the mineral surface which results in the release of metal ions from the solid surface (Brandl and Faramarzi, 2006; Vestola *et al.*, 2010).

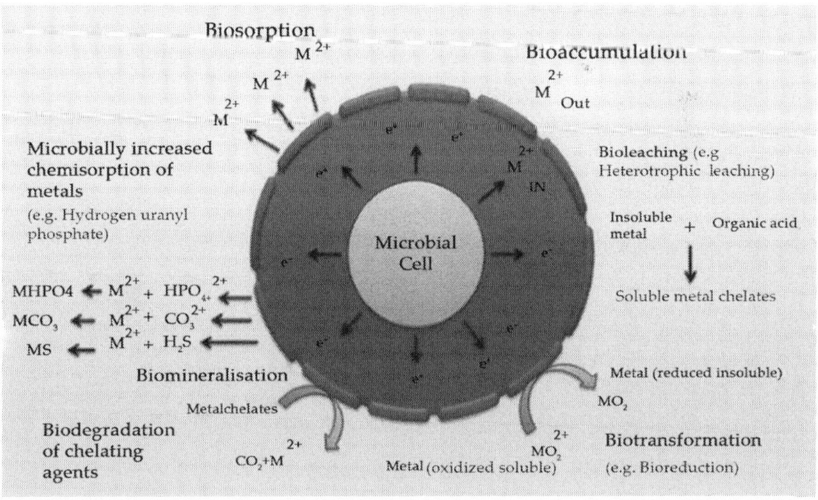

Figure 2.10: Microbial cell-interactions with metal and process of (bioaccumulation, bioleaching, biotransformation, biodegradation and biosorption; M^{2+}-metal ions) in soil with plant root rhizosphere surface.

The PGPR interacts with heavy metals that it can change the availability of them in soil and following features which bacteria possess (1) release of chelating substances(2) acidifying the microenvironment (3) inducing modifications in redox potential (Whiting *et al*., 2001). The lower the interaction of bacteria with the plant due to poor water supply, lack of soil fertility, harsh conditions and impossible nutrient exchanges(Figure 2.10).

2.8.2 Chelating Agents

Chelating compounds are a chemical substance composed of metal ions whose molecules can form several bonds with single metal ions. The solubility of heavy metals in soil increased positively by chelating agents and significantly increased the accumulation process by plants (Luo *et al*., 2015). The chelating agent used as a micronutrient fertilizer and maintains their solubility with metal (Socha and Guerinot, 2014). Chelating assisted phytoremediation demonstrated as a possible treatment for remediation of heavy metal contaminated soil and sediment. The toxic heavy metals exceed with a high amount of concentration then it certain threshold inside the cells to active metabolic process for the production of chelating substances (Oliva *et al*., 2012; Viehweger, 2014). Chelate agent contributes to metal degradation or detoxification by reducing the concentration of free metal in the cytosol (Soudek *et al*., 2014).

Ethylenediaminetetraacetic acid (EDTA) is appropriate for the phytostabilization of Pb-contaminated environments (Ullah *et al*., 2014). Previous reports suggest that a great number of chelating agents have been used for enhanced phytoextraction (EDTA, citric acid and malic acid). The introduction of EDTA and citric acid as a chelator is considered as important in chelation therapy (Komarek *et al*., 2007). Ethylenediamine tetraacetic acid was reported as a chelating agent for the enhancement in the phytoextraction process (Blaylock *et al*., 1997). It was recognized as the most efficient that increases metal mobility to uptake by plants for large scale field application (Tandy *et al*., 2004; Ghnaya *et al*., 2013). Firstly, chelate agent can absorb metals from the soil matrix the mobilized metals uptake by plant roots with the help of mobilizing activity of microorganisms.

Chelant agents in low amounts can enhance the phytoextraction of Pb (Shen *et al*., 2002). A combination of a chelating agent can also be very effective to improve the metal phytoextraction efficiency by lowering the pH of the soil with one type of combination of two chelants chemicals (Blaylock *et al*., 1997). The combination of EDTA and (S,S)-EDDS led to a higher level of efficiency that could be obtained by the application of either chelant alone performed in the phytoextraction of Zn, Cu, Pb and Cd (Luo *et al.*, 2015). Chelating agents are the most prevalent removal agents for high biomass production and are easily harvested (Mathis and Kayser, 2001). Interestingly applying one large dose of chelant often results that extraction of more toxic metals (Finzgar and Lestan, 2007). Studies suggested that when applying 10 mM EDTA of 1kg soil is used to remove Pb (Salido *et al.*, 2003). The field demonstrations are an important decrease in soil Pb concentration over two years at two sites in the United States (Robinson *et al.*, 2003). Lead (Pb) appears to move from roots to shoots as a metal-complex in the xylem (Huang *et al.*, 1997) via the transpiration stream (Blaylock *et al.*, 1997). Applying chelant agents in a low amount of dose can enhanced the phytoextraction of Pb (Shen *et al.*, 2002). Phytoremediation technologies have been developed commerciallyby*Alyssum*Nickle (Ni)-hyperaccumulator species to remove Ni from soils after treated with chelates (Broadhurst *et al.*, 2004; Akhter *et al.*, 2014).

2.9 ROLE OF PLANT NUTRITION

Heavy metals such as Cu, Zn, Al, Cd, Ni, Hg and As have negative effects on macronutrient and micronutrients (Hossain *et al.*, 2014). Calcium ions serve as a secondary messenger in signal transduction and movement in plants is unidirectional. Calcium is unable to recycle after deposited in the leaves (Hanger, 1979). Heavy metals bind to all calcium binding sites on the cell surface, at low pH (4.5) it works with calcium absorption and uptake calcium by roots (Roy *et al.*, 1988, Hossain *et al.*, 2014). A high amount of heavy metals exposure results the inhibition of calcium (Ribeiro *et al.*, 2013; Hossain *et al.*, 2014). Aluminum ions interfere with the action of Guanosine 5' triphosphate binding protein as well as inhibit calcium ion uptake by binding verapamil-specific channel (Rengel and Elliott, 1992). The uptake of calcium ions decreased in beech plants due to the combination of nitrogen and aluminum at high concentrations (Bengtsson *et al.*, 1994).

Potassium channel influx inhibits due to the toxicity of heavy metals like Aluminium. Active pathway involvement of uptake potassium is also inhibited by a high concentration of aluminum (Ribeiro *et al.*, 2013; Hossain *et al.*, 2014). At low pH 4.5, the concentration of potassium decreases, when they are treated with heavy metals (Moustakes *et al.*, 1995). Potassium ions decrease in the roots cell as well as guard cell due to the aluminum toxicity which causes by blocking the channels at the cytoplasmic site of plasma membrane (Liu and Luan, 2001). Somewhat significantly increase potassium ion from barley roots under aluminum stress (Kasai *et al.*, 1992). Potassium ions increase in root and shoot with an increase in aluminum concentration in pine trees. Durum wheat is more tolerant against aluminum toxicity and also decrease in potassium ion concentration with winter wheat (Zsoldos *et al.*, 2000).

Magnesium affected by heavy metals stresses more than other nutrient parameters of the plant. Plant treated with different concentrations of heavy metals and the level of magnesium decreases at pH (4.5). Magnesium uptake trough roots much stronger than calcium uptake (Bose *et al.*, 2013). Concentration decreased magnesium ions in roots and shoots by increasing the concentration of heavy metals (Huang and Bachelard, 1993). Iron concentration decreases at a pH of 4.5 when they are treated with different concentrations of heavy metals and also significantly much- affected root growth (Moustakas *et al.*, 1995). Plants exposed to a higher amount of heavy metals results to decrease the amount of lipoxygenase enzyme, ROS, antioxidant enzymes and also decline physiological effect (Wang and Yang, 2005). The soil has a high amount of heavy metals stress then they have a low amount of phosphorus and positively correlated with aluminum (Liu and Luan, 2001). The concentration of metals is present in low amount then it is required to increase such micronutrients and also macronutrients but if they are present in more than required amounts, they result to cause toxicity and also decrease the level of phosphorus (Nichol *et al.*, 1993, Cumming *et al.*, 1986).

2.10 FIELD EXPERIMENT IN PHYTOREMEDIATION TECHNOLOGY

Uranium (U) was accumulated in *F. antipyretica* and *Callitrichaceae* by aquatic plants and seems to be suggested for the phytoremediation process in field conditions (Paulo *et al.*, 2014). Researchers have investigated the efficiency of Pb, Zn, Cu and Cd removal with the help of chelators and $FeCl_3$ amendments by *Sedum alfredii* and *Zea mays*. The chelators were favorable to the removal of contaminants from the topsoil in a field experiment (Guo *et al.*, 2016). Previous work has investigated that silicon (Si) influences the improvement of Cd toxicity by peas in the field experiment. The plant growth, total protein and translocation noticeably were increased by the addition of Si under Cd stress. Previous studies suggested first evidence on the beneficial effect of Si on Cd toxicity in pea plants in field conditions

(Rahmana *et al*., 2017). There are few field-based observations on fluoride uptake by paddy, little information is available on the pattern of uptake and transfer of F in the paddy plant parts in a controlled condition. F accumulation in soil and vegetation in the vicinity of brick fields was previously reported (Jha *et al*., 2012).

2.11 EXPERIMENTAL PLANT (*PROSOPIS JULIFLORA*)

Prosopis juliflora is an appropriate candidate for phytoremediation of F polluted soils (Saini *et al*., 2012). *P. juliflora* is an evergreen tree native to South America, Central America and the Caribbean. In the United States, it is well known as mesquite.*P. juliflora* is a Fabaceae (leguminous): sub-family (Mimosoideae), perennial phreatophyte tree, extensively distributed in arid and semi-arid regions of the world and commonly known as Vilayati babool in India. This is also naturally grown in F widespread areas of Rajasthan (India) without showing any morphological distortion. It is tolerant of very high temperatures (like 48°C). The amount of nutrient content in *P. juliflora* is known to be of the highest value among the different leguminous plants.*P. juliflora* is highly admired for animal feed. It is fast- growing, nitrogen-fixing and tolerant to arid conditions and saline soils. This tree can grow in different areas, as well as those with saline, alkaline, sandy, and rocky soils and the roots enter to large depths in the soil.

3

Materials and Methods

CONTENTS

3.1 MATERIALS

3.1.1 Chemicals

Various chemicals and reagents of analytical molecular grade were obtained from Merck (India/Germany), Qualigens, Hi -Media, (India), Sigma (U.S.A). All solutions were prepared using Milli-Q water (Mill-Q, Integral A-10 System, France). All reagents were used of high purity grade.

3.1.2 Chelate Agents

Ethylenediamine tetraacetic acid (EDTA) and citric acid (CA) were collected from Hi -media (India).

3.1.3 Bacterial Strains

Two rhizobacterial strains (*Pseudomonas fluorescens* and *Pseudomonas aeruginosa*) were obtained in lyophilized form from MTCC, Chandigarh, India.

3.1.4 Soil Samples

Random soil samples were collected from the rhizosphere region of *Prosopis juliflora* were screened from different sites at the Banasthali Vidyapith for the isolation and characterization of plant-growth- promoting bacteria under F contaminated soil during winter (December) seasons. Soil samples were collected from a depth of 15-30 cm nearby areas of Braham mandir, Botanical garden and Krishi vigyan Kendra block of Banasthali Vidyapith Figure 1 (26° 60'N 75°54'E) Tonk (Rajasthan), India. Samples were brought to the soil testing laboratory for further chemical analysis. GPS points were taken keeping in view that five soil sample locations are representing a single site of Banasthali. GPS survey helped to plot the latitudinal and longitudinal information in real- world coordinate system. The GPS enabled data containing the positional location of the source of sample collection was further utilized to analyze the soil quality parameters in geo-spatial distribution patterns using geo- statistical tools.

3.1.5 Plant Materials

Seeds of *Prosopis juliflora* using in this study were collected from the Central Arid Zone Research Institute (CAZRI), Jodhpur, Rajasthan (India).

3.2 METHODS

3.2.1 Soil Processing and Physico-chemical Analysis

Collected soil samples for the isolation of bacteria were screened for physico-chemical properties. Soil moisture content was determined by drying sieved soil sample at 115° C for 48 hours as the difference in the mass before and after drying. These were done in Soil Science laboratory KVK Banasthali Vidyapith, Newai.

3.2.2 Determination of Soil pH

The pH was determined using the process of (Allen, 1989). The soil amount (2.5 g) soil was taken and passed through a 2 mm sieve. A 6.25 ml 0.01M $CaCl_2$ was added. It was well-mixed and kept overnight for about 14 hrs. The supernatant was centrifuged at 5000 rpm for 5 min and measured by pH meter (Systronics).

3.2.3 Determination of Electrical Conductivity

The electrical conductivity (EC) was measure in dS/m using 'Systronics' Conductivity meter. The instrument was calibrated and set for 0.01M KCl solution (1413 dS/m at 25°C). Similarly 1 dS/m = 0.65 ppm cation (Tandon,2002).

3.2.4 Carboncontent

Standard potassium dichromate solution (0.1667 M) = 49.04 g of K_2CrO_7 (dried at 105°C for 2 hours) in distilled water and dilute to 1 liter.

Ferrous sulfate or ferrous ammonium sulfate solution (0.5 M = 0.5 N) = 140 g of $FeSO_4.7H_2O$ or 196.10 g $FeSO_4$. $(NH_4)_2SO_4.6H_2O$ in 800 ml water, add 100 ml of concentrated H_2SO_4 cool and dilute to 1 liter.

Diphenylamine indicator = dissolve 0.5 g in a mixture of 20 ml water and add 100 ml of concentrated H_2SO_4.

Sulfuric acid = Concentrate less than 96%. If a high amount of chloride is present in the samples add silence sulfatesulfate (Ag_2SO_4) at the rate of 15 g/liter to the acid.

Ortho-phosphoric acid = 85% (85 ml orthophosphoric acid and 15 ml distilled water).

Soil samples were used for the determination of carbon percent by(Walkley and Black, 1934). Each sample was oven dry and weighed to obtain dry soil. Subsequently, the samples were oven- dried at 80° C for 5 days. Take 1 g of the soil sample in a 500 ml flask. Make two blanks for standardized with $FeSO_4$ or $FeSO_4$ $(NH_4)_2SO_4$ solution. Add 10 ml dichromate solution in a soil sample and gently stirring with a stirrer. Add 20 ml H_2SO_4 into the suspension and allow the flask to stand for 30 mins. After 30 mins add 200 ml distilled water, 10 ml of phosphoric acid and 500 µl of diphenylamine indicator and after this titrate this suspension with ferrous ammonium sulfatesulfate till the color flashes from blue to green. The C concentration was calculated as:

1. Blank- sample = A
2. $\frac{A \times 0.003 \times 100}{2 \times 1 \text{ g (soil)}} = B$ (organic carbon)
3. Actual carbon = $B \times 1.3 = C$
4. Organic matter = $C \times 1.724 = D$

3.2.5 Nitrogen Content

Total nitrogen was estimated by the micro-Kjeldahl method (Kjeltec Auto analyzer 1030 was used for N estimation).

3.2.5.1 Digestion

The plant samples (100 mg)were taken and transferred to a digestion flask. Add 100 mg of catalyst mixture and 5.0 ml concentrated sulfuric acid was added to it. Digestion process was carried out on a digestion unit till the solution became colorless. The contents of the digestion flask were transferred to a 25 ml volumetric flask up to the volume to 25 ml using distilled water.

3.2.5.2 Distillation

Take five ml of digested samples and put into the distillation unit and distillation carried out by adding 10 ml of 40% (w/v) NaOH. The released ammonia was a trap in 25 ml of 1 % boric acid solution containing a mixed indicator at the other end of the condenser tube. When all the ammonia was trapped, it was back titrated with 0.01 N HCl solutions. A blank sample containing distilled water was digested and distilled in a similar manner as explained above. The titer volume obtained for blank was used for correcting the nitrogen content in the distilled water.

$$\% N = \frac{\text{ml of HCl in sample determination}}{\text{ml of HCl for blank sample} \times \text{weight of the sample}} \times \text{Normality of HCl} \times 14.607$$

3.2.5.3 Soil Nitrogen

Potassium permanganate ($KMnO_4$) solution (0.32%) = 3.2 g in 1000 ml of distilled water.

Sodium hydroxide (NaOH) solution (2.5%) = 25 g NaOH in 1000 ml distilled water and mix well store in the plastic container.

Liquid paraffin (extrapure).

Sulfuric acid (0.02N) (N/50) for standard.

Mixed indicator = 0.07 gm methyl red with 0.1 g bromocresol green in 100 ml of 95% ethanol (alcohol).

Boric acid indicator solution = Take 20 g pure boric acid (H_3BO_3) in about 700 ml of hot water transfer the cooled solution to a 1-liter volumetric flask containing 200 ml of ethanol % 20 ml of mixed indicator solution. After mixing the contents of the flask add approx 0.5N NaOH continuously until the color is reddish- purple. Then dilute the solution to volume with water and mix it thoroughly. Soil samples were used for the determination of nitrogen content by (Subbiah and Bajaj, 1962). Take 20 g of soil in an 800 ml dry flask and add 20 ml of distilled water and mix properly.Add 1 ml of liquid paraffin and add 100 ml each of 0.32% $KMnO_4$& 2.5% NaOH. Distilled the contents in a Kjeldahl assembly at a steady rate & collect the liberated ammonia in a flask (250 ml) containing 20 ml of boric acid solution with mixed indicator. Absorption of ammonia causes the pink color of the boric acid solution to green color. Suspension of green color about 100 ml of distilled is to be collected in about 30 minutes. Titrate the contents with 0.02 N H_2SO_4 to the original shade (pink). Take blank without soil is to be made for the final calculations. The N concentration was calculated as:

1) Sample – Blank = R
2) N = R × 31.36 (kg/ha)

3.2.6 Determination of Available Phosphorus in Soil Samples

Sodium bicarbonate (Olsen's reagent) (0.5M $NaHCO_3$, pH 8.5)= Take 42 g $NaHCO_3$ in water make up to 1000 ml adjust pH 8.5 with (1M NaOH- 4 g NaOH/ 100 ml) solution. Take 10-15 ml NaOH solution is required for 1 liter of $NaHCO_3$.

Reagent A:

(a) 12 g ammonium molybdate $(NH_4)_6$ $MO_7O_{24}.4H_2O)$ in 250 ml of distilled water.

(b) 0.290 g antimony potassium tartrate (K $(SbO)C_4H_4O_6$. ½ H_2O in 100 ml distilled water.

(c) Add these two solutions to 1000 ml of 2.5 M H_2SO_4 mix thoroughly and make it to 2000 ml. Store in a pyrex glass bottle in a dark and cool place.

***Reagent B*: Freshly prepared**

a) 1.056 g ascorbic acid $(C_6H_8O_6)$ in 200 ml reagent A and mix. This does not keep for more than 24 hours at room temperature. Freshly prepared when it is required.

b) Sulfuric acid (2.5 M) = Dilute 140 ml of concentration of H_2SO_4 to 1 liter.

Standard stock (P) solution = Weigh 0.439 g of potassiumdihydrogen orthophosphate (KH_2PO_4) AR grade dried in an oven at 60°C for 1 hour and cool in desiccator in half a liter of distilled water. Add 25 ml of 7N H_2SO_4& make to one liter with distilled water. This gives a 100 ppm P standard stock solution. From this 2 ppm solution is add by diluting it 50 times. The samples were first extracted with Olsen's reagent and then available P was determined according to the method of Olsen(1982). Air-dried and 1 mm sieved soil samples (2.5 g) were taken in 250 ml (chromic acid- treated) Erlenmeyer flask and a pinch of P-free activated charcoal was added to it. Then 50 ml of Olsen's reagent (soil to solution ratio of 1:20) was added and kept on the shaker for 30 min. Similarly, a blank was run without soil. All the sets were taken in triplicate. The samples were filtered through Whatman no. 42 filter paper into a clean dry beaker. Flasks were shaken immediately before pouring the suspension into a funnel. Take five ml of the extract was taken and acidified to pH 5 with 2.5 M H_2SO_4. The volume of the samples was made to 20 ml with double distilled water and then 4 ml of reagent B was added. After 20 min the intensity of blue color was read at 840 nm using a spectrophotometer. The P concentration was calculated as:

$$\text{Available P} = \frac{\text{R}\times\text{volume made at color develop}\times\text{volume of extract (ml)}}{\text{the volume of aliquot}\times\text{weight of soil (g)}}$$

Where R stands for the concentration of P (mg/l) in the sample obtained from X-axis against the spectrophotometer reading.

3.2.6.1 Preparation of standard curve for determination of available P in soil samples

Different amounts (0.5, 1.0, 2.0, 3.0, 4.0, 5.0, 6.0, 7.0, 8.0 and 10.0 ml) of 2 ppm Phosphorus solution was pipette out in a 50 ml tube and 5 ml of Olsen's reagent was added to it. The solution was acidified to pH 5.0 with 2.5 M H_2SO_4. The bubbles were allowed to escape and the volume was made up to 20 ml with double distilled water. Take four ml of reagent B was added to it, vortexed and left undisturbed for 20 minutes. The intensity of blue color was measured at 840 nm using a spectrophotometer (Systronics UV-VIS Spectrophotometer 119, Naroda, Ahmedabad, India). Similarly, a blank was run without standard phosphate. The graph was plotted between the concentration of the standard phosphate solution and their respective absorbance values.

3.2.7 Determination of Potassium

Ammonium acetate (extracting solution) (1.0N, pH 7) = Take 700 ml of distilled water and add 57 ml of glacial acetic acid (CH_3COOH) and then 69 ml of concentrated ammonium hydroxide (NH_4OH). Dilute to a volume of 900 ml and adjust pH to 7 by the addition of more of 3 N NH_4OH and make up to 1 liter. Store in pyrex or polyprophylene bottle. Alternative dissolve 154 g ammonium acetate (CH_3COONH_4) in water and dilute to 1.8 liters. Mix thoroughly and adjust the pH to 7 with dilute NH_4OH or CH_3COOH as required and make it to 2 liters.

Potassium chloride standard solution (KCl) = Make a stock solution of 1000 ppm K by dissolving 1.908 g of AR grade potassium chloride (dried at 60°C) in distilled water and diluting up to 1 liter. Prepare a 100 ppm K standard by diluting 100 ml of 1000 ppm stock solution to 1 liter with the extracting solution. The potassium from soil samples was determined by a flame photometer (Schollenberger and Simon, 1945). Take 5 g of soil in a 150 ml conical flask and add 25 ml of neutral ammonium acetate (pH 7). Shake on a reciprocating shaker for 30 minutes and filter thoroughly. Take suspension for reading by flame photometer. Take0, 5, 10, 15 and 20 ml of 100 ppm k solution into 100 ml volumetric flask for standard. Draw a standard curve by plotting the flame photometer readings against K concentration. The K concentration was calculated as:

$$\text{Available K (kg/ha)} = \frac{\text{R} \times \text{Volume of extract} \times 2.24}{\text{Wt. of soil sample}}$$

Where R is ppm of K in the extract (obtained from the std curve)

3.2.8 Determination of calcium and magnesium

3.2.8.1 EDTA and titration methods

Standard calcium solution (0.01 N): Weigh 0.5005 g of calcium carbonate dried at 150° C in 1 liter flask. Add 200 ml of water and 150 ml of 1 N HCl slowly and with shaking. Dilute to 1 liter.

EDTA solution: Dissolve 2 g of EDTA (disodium salts) in 1 liter distilled water.

Sodium hydroxide (10%): 10 g sodium hydroxide in 100 ml distilled water.

Buffer solutions: Dissolve 67.5 g ammonium chloride in 400 ml of water, add 570 ml of concentrated ammonium hydroxide and dilute to 1 liter.

Hydroxylamine-hydrochloride aqueous solutions (5%): 5 g in 100 ml of distilled water freshly prepared each week.

Potassium hexacyanoferrate aqueous solution (4%):4 g in 100 ml distilled water.

Potassium cyanide aqueous solution (1%):1 g in 100 ml distilled water.

Triethanol amine and Calcon solution: Dissolve 0.2 g in 50 ml methanol prepared fresh every 2 weeks.Eriochrome black T solutions: Dissolve 0.2 g in 50 ml methanol prepared fresh every 2 weeks.

Standardization was done by EDTA solutions for calcium and magnesium determination. Pipette 5 ml of standard calcium solution into a 100 ml beaker. Dilute to 10 ml and add 50 ml of ammonium chloride- ammonium hydroxide buffer solution. Add 10 drops each of potassium cyanide, hydroxylamine-hydrochoride, potassiumhexacyanoferrate, triethanolamine and eriochrome black T solution. Place the beaker on a magnetic stirring plate and start stirring. Prepare a blank solution in exactly the same manner, taking 5 ml of water instead of calcium solution. Usually the blank solution was blue in color, but if not, it should be titrated with EDTA solution until blue and the blank titer value noted. Keep the blue blank solutions alongside the standard calcium solutions and titrate the standard with EDTA solution, stirring all the time to permanent blue color matching the blank. Dilute the blank with water now and again to equalize the volumes of the two solutions as the titration proceeds. Repeat the standardization using the calcon as the indicator. In this case, add to dilute 10 drops each of potassium cyanide, hydroxyl amine-hydroxile and triethanol amine solution. Add 2.5 ml of sodium hydroxide and 1 ml of calcon solution. Prepare a reagent blank and titrate both solutions with EDTA solutions until blue.The calcium and magnesium analyzed by the titration method (Cheng and Bray, 1995). Take 5 g of soil in 150 ml flask and add 25 ml of neutral ammonium acetate buffer (Same as given in potassium estimation method). Shake for 5 minutes and filter immediately filter through Whatman no. 1 filter paper. Pipette an aliquot of the extract (5 ml) into a 100 ml beaker and dilute to 10 ml. Add 10 drops each of potassiumcyanide, hydroxyl amine-hydroxile and triethanol amine solution. Add 2.5 ml of sodium hydroxide and 1 ml of calcon solution. Titrate with EDTA until blue.

3.2.8.2 Magnesium

Magnesium is calculated from the difference between the calcium+magnesium and the calcium determinations(Ca+ Mg) – Ca). The Ca and Mg concentration were calculated as:

$$\text{Ca or (Ca+Mg), meq/liter} = \frac{T\times \text{ normality of EDTA} \times 1000}{\text{aliquot (ml) taken}}$$

Where T = volume in ml of standard EDTA used in titration

or

$$\text{Ca or (Ca + Mg) meq/100 g soil} = \frac{100 \times \text{Extract volume (ml)}}{\text{soil wt (g)} \times 1000} \times \text{meq Ca or (Ca+Mg)/liter}$$

3.2.9 Determination of Sulfursulfur

Calcium chloride dihydrate (0.15%): Dissolve 1.5 g calcium chloride in 1 liter distilled water.

Barium chloride dihydrate crystals: Take 0.25 g.

Conditioning reagent: Dissolve 75 g sodium chloride in 275 ml of distilled water in a 500 ml flask. Stirring with a magnetic stirring bar, add 30 ml concentrated HCl, 100 ml absolute ethanol and 50 ml of glycerol. Rinse glycerol into the flask. Continue stirring until the NaCl dissolves. Remove the stirring bar and make up the volume with distilled water.

Standard sulfatesulfate solution: Dissolve 0.54 g dipotassium sulfatesulfate in 1 liter distilled water. This contains 100 microgram sulfur/ml (100 ppm). Dilute these 10 times to obtain 10 ppm solutions.

The S content was determined by the method of Tondon, (2002). Take 5 g of soil into a 150 ml flask.Add 25 ml of 0.15% calcium chloride solution. After this, shake for 30 mins and filter through Whatman no. 42. Taken 10 ml filtrate extract in 25 ml flask and add distilled water thus bringing the volume up to 20 ml. Add 2.5 ml of conditioning reagent and 0.25 g barium chloride crystals with a spatula to the standards and extract and make the volume to 25 ml and shake the flasks for 1 min each. After 3 min, measure the turbidity in a colorimeter at 340 nm on a spectrophotometer. The S concentration was calculated as:

$$\text{Available sulfur } (mgkg^{-1}) = \frac{R\times \text{ volume made at turbidity development}}{\text{Volume of aliquot } \times \text{ weight of soil}} \times \text{volume of extract}$$

Where R = concentration of sulfur (mg/liter) in the sample obtained from the X- axis against the reading.

3.2.9.1 Preparations of standard curve for sulfur

Take 0, 2.5, 5, 7.5, 10, 12.5 and 15 ml of the working standard sulfate solutions into a series of 25 ml flask to obtain 0, 1, 2, 3, 4, 5 and 6 ppm sulfur respectively. Proceed to develop turbidity as described below for sample aliquots. Read the color intensity for each standard after adjusting blank to zero and prepare the curve by plotting reading against sulfur concentrations

3.2.10 Available Micronutrient in Soil

The diethylene triaminepenta acetic acid (DTPA) extractable Iron (Fe), Manganese(Mn), Zinc (Zn) and copper (Cu)were estimated by using atomic absorption spectrophotometer (AAS) (Lindsay, 1978).

DTPA = 0.005 M (393.35 formula weight)

$CaCl_2.2HO_2$ =0.01 M solution

Tri Ethanol Amine (TEA) = 0.1 M solution

To prepare 1 liter of DTPA extracting solution is dissolved in 13.1 ml reagent grade TEA, 1.96 g DTPA (AR grade) and 1.47 g of $CaCl_2$ in 100 ml of glass-distilled water. Allow some time for the DTPA to dissolve and dilute to approximately 900 ml. adjust the pH 7.3 with 1N HCl. Take 10 g of soil in 150 ml flasks and add 25 ml of DTPA extracting solution. Shake for 2 hrs at a speed of 120 cycles per minute. Filter the suspension through Whatman no. 42 filter paper. Keep the filtrate in polypropylene bottles for the analysis of Zn, Cu, Mn and Fe with AAS. Analyse the samples as described above using a standard curve. When samples need dilution before measurements, they should be diluted with DTPA solution to maintain a constant matrix.

3.2.10.1 Standard of zinc

Dissolve 0.43 g$ZnSO_4.7H_2O$ in 200 ml of glass distilled water in a beaker and add 5 ml of H_2SO_4 for getting 100 µg Zn/ml (100 ppm = mg kg^{-1} Zn) solutions. Diluted solutions with DTPA of 1, 2, 3, 4 and 6 ml prepared using the stock solution (100 ppm of Zn/ml).

3.2.10.2 Standard of iron

Dissolve 0.70 g ferrous ammoniumsulfate($(NH_4)_2SO_4FeSO_4.6H_2O$)in 300 ml of glass distilled water in a beaker and add 5 ml of H_2SO_4 for getting (100 ppm = mg kg^{-1}Fe/ml) solutions. Diluted solutions with DTPA of 1, 2, 3, 4 and 6 ml prepared using the stock solution (100 ppm of Fe/ml).

3.2.10.3 Standard of manganese

Dissolve 0.28 g potassium permanganate ($KMnO_4$) in 300 ml of glass distilled water in a beaker. Add 20 ml of concentrated H_2SO_4 warm to about 600°C and add oxalic acid solution dropwise to make the solution colorless. Transfer to 1 liter measuring flask and make volume to the mark to get a standard solution to 100 µg Mn/ml (100 ppm = mg kg^{-1} Mn). For preparing working standard transfer 1, 2, 3, 4 and 6 ml of stock solution (100 ppm Mn/ml) and dilute each to mark with DTPA extracting solution.

3.2.10.4 Standard of copper

Dissolve 0.39 g copper sulfate ($CuSO_4.5H_2O$) in 400 ml of glass distilled water in a beaker and add 5 ml of H_2SO_4 for getting (100 ppm Cu/ml) solutions. Diluted solutions with DTPA of 1, 2, 3, 4 and 6 ml prepared using stock solution (100 ppm = mg kg^{-1} Cu/ml).

3.2.11 Identification of isolated bacteria

3.2.11.1 Morphological identification of isolated bacteria

The morphological observations were done on colony characteristics like size, shape and color (Chao *et al*., 2013).

3.2.11.2 Identification of isolated bacteria by biochemical parameters

Isolated plant-growth promoting bacteria were characterized by biochemical parameters using methodology of Krieg (1984).

3.2.11.3 Gram staining

A thick smear of test culture was prepare on a slide and air dried. The smear was flooded with crystal violet for 30 sec followed by washing with tap water. Gram's iodine was added for 30 sec and washed with alcohol for 5-10 sec. Saffraninewere added for 30 sec, followed by washing with tap water. The slides were observed under the microscope. Violet color indicated positive reaction, whereas pink color indicated negative reaction.

3.2.11.4 Catalase

A loopful of culture from freshly grown slant of tryptone yeast extract was taken out and placed on the slide. A drop of H_2O_2 (3 %) was added to culture. Production of bubbles indicated presence of catalase.

3.2.11.5 Hydrogensulfide production test and motility

Stab inoculation in SIM (sulfide indole motility) medium was prepareaseptically to sterile labeled tubes. The inoculated tube was incubated at 37°C for 24-48 hours and the results were determined. The medium color changes from its normal color to black precipitateon the medium.

3.2.11.6 Sugar fermentation

Phenol red triptocase broths with Durham tubes were prepared using different sugars as carbon source. The cultures were inoculated in the tubes and incubated at 37°C for 48 h. Presence of yellow color indicated acid production and presence of bubbles in Durham tubes was indicative of gas production.

3.2.11.7 Nitrate reduction and Oxidase test

Nitrate broth was inoculated with test cultured and incubated at 37°C for 48 h. To test the presence of nitrite, 2 ml nitrite reagent and 1 ml of a-napthylamine reagent was added to the tubes and development of Pink color indicated nitrate reduction. To test complete denitrification a pinch of zinc dust was added to the tubes, reappearance of pink color showed complete denitrification. The cultures from freshly grown slant of nutrient agar were taken for oxidase test and add 1% (100 μl) tetramethyl-p-phenylenediamine dihydrochloride solution. Observe the area for a color change to deep blue with in 10 seconds which shows positive results.

3.2.11.8 Voges Proskauer test (V-P test)

MR-VP broth was inoculate with test cultures and incubated at 37°C for 48 hr. One ml of the culture broth was removed into a clean test tube and 0.6 ml of ethanolic ∞-naphthol (5%) was added and mixed well. Then 0.3 ml of 40% KOH was added and tubes were shaken. Observations were recorded for the appearance of red color within 5 minutes.

3.2.11.9 Carbohydrate utilization test

The carbohydrate utilization test was done with the help of Kit KB009 Hi Carbo™ Kit (KB009A, B and C). It was used to study the biochemical test of a wide varietyof the unknown organisms. The tests were done for utilization of 35 carbohydratetests viz Lactose, Xylose, Maltose, Fructose, Dextrose, Galactose, Raffinose, Trehalose, Melibiose, Sucrose, L-Arabinose, Mannose, Inulin, Sodium gluconate, Glycerol,Salicin, Dulcitol, Inositol, Sorbitol, Mannitol, Adonitol, Arabitol, Erythritol, Alpha-methyl-D-glucoside, Rhamnose, Cellobiose, Melezitose, Alpha-methyl-D-Mannoside, Xylitol, ONPG, Esculin, D-Arabinose, Citrate, Malonate, Sorbose and 1 control.

The kit contains part A, B and C each having 12 carbohydrate utilization tests and C containing 11 sugars and 1 control. Pure culture organisms identified and incubated at 37°C for 4-6 hours. Open the Kit aseptically. Peel off the sealing foil. Put 50 μl inoculate each well and keep for incubation at 37°C for 18-24 hours. The red color was change to yellow color due to acid production if the test is positive and medium remains red in color if the test is negative.

3.2.12 Molecular characterization of isolated and selected bacteria

Selected isolates were characterized by 16S rDNA gene sequence technique (Wintzingerode *et al.*, 1997).The DNA was extracted and amplification of 16S rDNA gene was performed for further analysis. The primers were used for amplification such as forward primer 27FP1 (5'- AGAGTTTGATCCTGGCTCAG-3') and reverse primer 1492RP2

(5'-ACGGCTACCTTGTTACGACTT- 3'). The 16S rDNA gene sequences were identified by direct sequencing PCR-base method using dye fluorescent terminator method. Finally, the products were analyzed in ABI 3730 XL capillary DNA sequencer (which have 50 cm capillary).

The 16S rDNA gene sequences were aligned and using BLAST for analysis EzTaxon server. Phylogenetic analyses were done using MEGA version 5 (Tamura et al., 2011) after multiple alignments of the data from CLUSTAL W. Clustering was performed by neighbor-joining method (Saitou and Nei, 1987).

The consensus bootstrap concrete by 1000 replicates for represents history of the taxa (Felsenstein, 1985). The replicate trees percentage was associated taxa together with clustered in the bootstrap test (1000 replicates). The statistical confidence was estimated by bootstrapping using 1,000 replications.

3.2.12.1 Isolated bacterial strains will be tested by enzymatic assay for plant growth promoting characteristics

3.2.12.1.1 Determination of 1-aminocyclopropane-1-carboxylate deaminase

This assay wasperformto utilize the nitrogen component using the method of (Saleh and Glick, 2001). Take sterilized *P. juliflora* seeds were allowed togerminate at 25°C for 48 h. Germinatedseeds were dipped for 30 min in 50 ml bacterial culturesgrown for 48 h in nutrient broth and transferred it to wet filter paper in petri dishes taking eight seeds per dish. The petri dishes incubated in dark at 30±7°C. The uninoculated nutrient broth served as control. The rootlength of seedlings was measure in cm after 14 days of incubation.

3.2.12.1.2 Ammonia production

The ammonia production test wasdetermine by the method of (Cappuccino and Sherman, 1992).To check the production of ammonia, bacterial isolates were grown in peptone water broth/ nutrient broth and incubated at 28 ± 2°C for four days. The accumulation of ammonia was detected by addition of Nessler's reagent 1 ml per tube. A faint yellow color indicated small amount of ammonia

3.2.12.1.3 Hydrogen cyanide production

Hydrogen cyanide production assay were perform on bacterial isolates (Castric *et al.*, 1975). Takenutrient agar medium with 4.4 g L^{-1} of glycine and then sterilized. The medium transfer or poured into sterilized petri plates and streaked with 24 hr old bacterial strain. Take Whatman no.1 filter paper disc, dip in picric acid (0.05% solution in 2% sodium carbonate) and was placed in petri plate. Take one petri plate as a control which did not receive inoculums.

The petri plates were sealed with parafilm and keep these plates for incubation at 28 ± 2°C for 6 days. The color change of the filter paper (deep yellow to orange and then to orange brown and finally to dark brown) was positive. The negative test was remained unchanged after the growth of the bacteria.

3.2.12.1.4 Determination of indole acetic acid

Indole acetic acid (IAA) production by bacterial strains was performaccording to the method (Loper and Schroth, 1986). Take 1 ml of each bacterial strain culture inoculated in 50 ml of nutrient broth with tryptophan (1 and 3 mg ml^{-1}) or without tryptophan and incubated at 28 ± 2°C for 7 days. A 5 ml sample culture was removed from each flask after every 24 h and centrifuged at 3000 rpm for 30 min.

An aliquot of 2 ml of the supernatant was transferred to a fresh tube, to which 100 µl of 10 mM orthophosphoric acid and 4 ml of Salkowaski reagent (50 ml, 35% perchloric acid; 1 ml $FeCl_3$) were added. The incubation time was 25 min at room temperature and the development of pink color indicated IAA production.

3.2.12.1.5 Antibiotic sensitivity assay for F resistance

Isolated bacterial strains were tested for their sensitivity towards various antibiotics viz., (tetracycline, streptomycin, chloramphenicol, gentamycin and ampicillin)at concentration of 30 mcg $disc^{-1}$. Isolate bacteria show sensitivity to the respective antimicrobial disk sensitivity test by following the National Committee for Clinical Laboratory standards (NCCLS).

Different antibiotics discs were placed on the medium and incubated for 48 hrs. Plates having zones of inhibiton around the antibiotic discs showed that the strain was sensitive to the respective antibiotic.

3.2.13 Effect of the isolated bacterial strains on plant growth and F uptake

Stock F solution (1000 mg/l = 1000 ppm = 1000 mg kg^{-1}): 2.21 g NaF in 1 litre deionized water.

Total ionic strength adjustment buffer (TISAB): 58 ml glacial acetic acid+12 g sodium citrate dehydrate in 1 litre distilled water adjusted at the pH 5.2 by using 6 N NaOH.

The pot experiments were carried out the effect of strain on the plant growth and uptake of plants under F-contaminated soil. Seeds of *P. juliflora* were collected from the Central Arid Zone Research Institute (CAZRI) Jodhpur (Rajasthan) India. Before sowing, seeds were surface sterilized and soaked in distilled water for 12 hrs.Seeds of *P.juliflora* were surface-sterilized with 10% H_2SO_4 for 10 min and washed with sterile double distilled water. Seeds were germinated in plastic pots (15-cm height) with 1 kg of the sterilized testing soils without any amendments of microbes. Three pots were used for each treatment to avoid error (replication). Five surface-sterilized seeds were placed in each pot at a 1.0 cm depth. The inoculation, strains were grown in nutrient broth. The isolated *Clostridium sp.*was used as inoculums for the treatment.

The strain was inoculated in the amount of 10^8 CFUCFU $mL^{-1}kg^{-1}$and F concentrations for treatment was decided. Next, it was inoculated with test cultured and incubated at 37°C for 48 h.One ml of the culture was taken into a clean beaker and 9 ml of sterile water was added and mixed well in field soil (1 ml for 1 kg of soil in pot. Treatment has also given concentrations of F (25, 50, and 75 and 100 mg kg^{-1} NaF).

Uninoculation was taken as a control. There are three pots without F and *Clostridium sp.*was supplied with millipore water.Plants were removed from pots and the soil was removed from the root and harvested after 120 days, washed with tap water. The root and shoot length were directly measured with scale and the root, shoot fresh weights and root, shoot dry weights were also recorded before and after oven-dried at 70° for 72h.The following reagents were prepared for the determination of F.

3.2.13.1 Parameters studied

Following parameters were studied during the growth of plants and after their harvest.

3.2.13.2 Germination percentage

After 6 days of sowing, the germinated seeds were counted and the germination percentage was calculated by the following formula:

$$\% \text{ Germination} = \frac{\text{Number of seeds germinated}}{\text{Number of seeds sown}} \times 100$$

3.2.13.3 Root and shoot measurements

Root, shoot length and root, shoot dry weight and fresh weight (including leaves) were measured after 30, 60, 90 and 120 days after sowing *P. juliflora*. For measuring root and shoot dry weight, the samples were dried in an oven for 48 hours at 60°C.

3.2.13.4 Determination of F content in soil

The total F content in the soil was determined using the method of McQuaker and Gurney (1977). The collected soil samples were dried overnight at 80°C and these dried samples were grounded to fine powder. Approximately 0.5 g of powdered soil samples was taken and transferred to the 130 ml nickel crucible.

The sample was moistened slightly with distilled water. This was followed by addition of 6 ml of 16 N NaOH. The crucible was tapped slightly so as to uniformly disperse the samples in the NaOH solution.

The samples were then placed in oven set at 150°C for 1hr. For the formation of NaOH and sample fusion cake, the crucible was placed in muffle furnace till the temperature reached 300°C.

The temperature was raised to 600°C and the samples were fused at this temperature for 30 min. The samples were removed, allowed to cool, and then 10 ml of distilled water was added to the samples, slowly with stirring to adjust the pH to 8-9. The sample were filtered and transferred to 100 ml volumetric flasks and then were diluted with double distilled water to make up the final volume up to 100 ml and filtered through dry Whatman no. 40 filter paper.

Three independent blanks were also run simultaneously with the samples and then mixed to obtain a representative blank. Take 5 ml of the above extract, 5 ml of TISAB was added and mixed, and the F measurement was done by F ion- selective electrode using ORION 4 Star ion analyzer. The detection limit of method (LOD) was 0.05 mg L^{-1}.

3.2.14 Determination of the total soluble F content in the plant

Fluoride content was determined by the method described in 3.2.13.4. Soil-grown seedlings after a period of 30, 60, 90, 120 days of F treatment were harvested and dissected into different organs i.e., roots and shoots. Dried materials were grounded in a morter and pestle. Powdered sample (0.5 g) was transferred to a 130 ml nickel crucible. The detection limit of method LOD was 0.02 mg/l.

3.2.15 Determination of bioaccumulation factor (BF) and trans-locationFactor (TF)

Bioaccumulation factor and translocation factor were calculated by using method of (Zhao *etal*., 2003).

BF = [F concentration in shoot] / [F concentration in soil]

TF = [F concentration in shoot] / [F concentration in root]

3.2.16 Effect of microorganisms and chelating compounds on biochemical, nutrientcontent and (F) accumulation of hyperaccumulator plant *P.juliflora*

3.2.16.1Pot experiments

3.2.16.1.1Seed germination, microinoculation and chelating compounds in the soil for studying the effect of NaF on different parameters

P. juliflora seeds were collected from the Central Arid Zone Research Institute (CAZRI), Jodhpur (Rajasthan) India. Seeds were surface sterilized with 10% H_2SO_4 for 15 min and rinsed with millipore water. Seeds were germinated in plastic pots with 1 kg of sterilized soils. Each pot received six seeds that were placed at 1cm depth. Pots were rearranged in the greenhouse chamber. The F contaminated soil of pot experiment was performed artificially given NaF of different concentrations of 25, 50, and 75 and 100 mg kg^{-1} in soil. Another set of experiment was subjected to a different type of inoculation: PF 8904 (*Pseudomonas fluorescens*), and PA 1934 (*Pseudomonas aeruginosa*) strains obtained from Microbial Tissue Culture Collection (MTCC) Chandigarh together with F concentration. Three replicates were used for each F level inoculation type with and without microbial treatment.

Bacterial strain suspension (10^8 CFUCFU ml^{-1}) in nutrient broth was used for the inoculation, by spraying soil surfaces (Marques *et al.*, 2010), after 10 days of germination. To the control pots, 10 ml of sterile distilled millipore water was added. The chelates concentrations (5, 10, 15, 20 mM kg^{-1} soil) were applied, alongside with untreated (without EDTA and CA) control group. For each concentration of F in soil three independent sets of plants were maintained. Plants were harvested after 120 days, washed with tap water and deionized sterile water for further analysis. Plant biomass (roots and shoot length) was determined using the measuring scale.

3.2.16.1.2 Determination of photosynthetic pigments

Fresh experimental plant leaves were taken for the determination of photosynthetic pigments e.g. Total Chlorophyll (total Chl), Chlorophyll a (Chl a) and Chlorophyll b (Chl b). The leaves cut into tiny pieces and homogenized in 5 ml of chilled 100 % acetone in a pre-cooled mortar pestle until the powdered materials become completely non-green. The homogenate was centrifuged for 5 min at 3000 rpm at 4°C in a cooling centrifuge. The pellet was discarded and the supernatant was re-adjusted to 5 ml acetone. To 1.6 ml of the supernatant, 0.4 ml double distilled water was added. The chlorophyll content absorbance of the resulting supernatant was recorded at 645 and 663 nm using a double beam UV-Vis spectrophotometer(Arnon *et al.*, 1949) as given below:

$$\text{Total Chl (a and b) (mg/g)} = 20.2\,(A_{645}) + 8.02\,(A_{663})$$

$$\text{Chl a (mg/g)} = 12.7\,(A_{663}) - 2.69\,(A_{645})$$

$$\text{Chl b (mg/g)} = 22.9\,(A_{645}) - 4.68\,(A_{663})$$

The above formula expressed the content of the pigment in mg g^{-1} fw of the sample.

3.2.16.1.3 Anioxidant enzyme activities under varying concentration of NaF

3.2.16.1.3.1 Peroxidase activity

Take 1g of fresh leave with 3 ml of 0.1 M phosphate buffer (0.1M) pH 7.0 by grinding in a pre-cooled mortar and pestle. After the homogenate process was done, put the samples in a centrifuge at 10,000 rpm at 5°C for 15 min. Peroxidase (POD) activities were determined specifically with guaiacol (20 mM) at 436 nm following the method of (Putter *et al.*, 1974). Enzyme extract (0.1 ml) was added to the reaction mixture containing 0.05ml guaiacol solution and 0.03 ml hydrogen peroxidase solution (0.042%) in 3 ml phosphate buffer solution (pH 7.0). The solution was then mixed well and kept until the absorbance at 436 nm read 0.05 in the spectrophotometer. Time was then noted for the absorbance to increase by 0.1. The enzyme activity was calculated using the extinction co-efficient of guaiacol dehydrogenation products under the conditions specified.

3.2.16.1.3.2 Catalase activity

Take 1 g of fresh plant tissue and then 3 ml of 0.06 M phosphate buffer (0.06 M = 3.52 g) pH 7.0 by grinding in a pre-cooled mortar pestle. Centrifuge the homogenate samples at 10,000 rpm at 5°C for 15 min. A modified method of (Luck *et al.*, 1974) was employed for the assay of catalase (CAT). Enzyme extract (50 μl) was added to 3 ml of hydrogen peroxide (10% w/v) phosphate buffer (pH 7.0). The time required for the decrease in absorbance from 0.45 to 0.40 was noted. Enzyme solution containing hydrogen peroxide-free phosphate buffer was used as a control.

3.2.16.1.3.3 Superoxidase activity

The activity of SOD was assayed by monitoring the inhibition of photochemical reduction of nitro blue tetrazolium (NBT), according to Beauchamp and Fredrovich (1971). Enzyme extract was add to 3 ml of reaction mixture containing, 50 mM phosphate buffer (pH 7.8), 13 mM methionine, 75 mM NBT, 0.1 mM EDTA and 0.2 mM riboflavin. Riboflavin was added at the end and the tubes were shaken and placed 30 cm below a light source consisting of two 15 W fluorescent lamps. The reaction was started by switching on the light and was allow running for 15 min. The reaction was stopped by switching off the light and the tubes were covered with black cloth. The photo-reduction of NBT (production of blue formazan) was measured at 560 nm, and an inhibition curve was making against various volumes of the extract. One unit of SOD was defined as the amount of enzyme required to cause 50% inhibition of the reduction of NBTat 560 nm.

3.2.16.1.4 Determination of total protein content

The following reagents were used:

BSA solution 1mg/ml

Reagents A: 50 ml of 2% sodium carbonate mixed with 50 ml of 0.1 N NaOH solution(0.4 gm in 100 ml distilled water)

Reagent B: 10 ml of 1.56% copper sulfate solution mixed with 10 ml of 2.37% sodium potassium tartrate solution.

Reagent C: 100 ml of reagent A and 2 ml of reagent B was mixed.

Folin's-Ciocalteau reagent: 2 ml of Folin's reagent was mixed with 2 ml distilled water and this reagent was prepared freshly. The estimation of total proteins was done using the method of Lowry *et al.*, 1951.

3.2.16.1.4.1Extraction of total proteins

1 g tissue was taken with 0.1 M phosphate buffer at pH 7.0 and centrifuged at 15,000 rpm at 4°C. The pellet was discarded and supernatant obtained was used for further analysis.

3.2.16.1.4.2 Estimation of total proteins

100 μl of supernatant was taken in a test tube and volume was made up to 1 ml with distilled water.5 ml of Reagent C was added to the reaction mixture and was incubated for 10 min at room temperature. Then 0.5 ml of Folin's-ciocalteau reagent was added and then the reaction mixture was further incubated at room temperature for 30 min in dark. The blue color was developed in each sample and the absorbance was recorded at 660 nm using a UV-2450 spectrophotometer. BSA was used to make thc standard curve.

3.2.16.1.5 Determination of F

The total soluble F was determined by the method described in 3.2.13.4 section and bioaccumulation and translocation factors were also calculated by the method explained in 3.2.15 after treatment with microbes and chelates.

3.2.16.1.6 Processing and analysis of macro and micronutrients from plant organs (roots, shoots and leaves)

3.2.16.1.6.1 Mineral content of plant organs (Root, Shoot and Leaves)

The plants were harvested after 120 days of inoculation and analyzed for further experiments by the method described in 3.2.2-3.2.11 section. The whole plant was washed with distilled water, separate out into root and shoot and leaves, then oven-dried for 72 hr at 70°C to determine the contents of plant organs. The plant organs (root, shoot and leaf) were analyzed for Nitrogen-N, Phosphorus-P, Potassium-K, Iron-Fe, Manganese-Mn, Zinc-Zn and Cooper-Cu.

3.2.17 Effect of *Pseudomonas Fluorescens* and Ethylene Diamine Tetraacetic Acid on Biochemical, Mineral Contents and (F) Accumulation by*P.juliflora* in Field Experiment

3.2.17.1 Design of the test field site

The field study was designed on the basis of phytoextraction experiments conducted at the laboratory. The field soil was sandy, loam and saline in nature. Firstly, the test area soil was excavated and immediately divided into twenty seven plots (length × width =2m×1m=2m^2) (Figure 3.1), which were then bordered at the sides to prevent lateral flow between the plots. Soil wetland was constructed at an experimental site of 2m^2 (length×width) and 0.15 m depth Figure 3.2.On the basis of pot experiments, *P. fluorescens* was inoculated in the amount of 10^8 CFU mL^{-1}kg^{-1}and F concentrations for treatment was decided. Next, it was inoculated with test cultured and incubated at 37°C for 48 h. One ml of the culture was taken into a clean beaker and 9 ml of sterile water was added and mixed well in field soil (1 ml for 1 kg of soil in a pot (1 ml culture+9 ml D.W) and 100 ml culture+900 ml D.W for 100 kg of soil) (approx amount). Seeds of *P. juliflora* were collected from Central Arid Zone Research Institute (CAZRI) Jodhpur (Rajasthan) India. Before sowing, seeds were surface sterilized and soaked in distilled water for 12 hrs. The next day, seeds were washed and treated with tween 20 for 5 min. Seeds were then again washed under tap water; sodium hypochlorite (2 ml) was applied for 5 min and finally rinsed under tap water for further use. Six seeds were sowed on one line in soil bed at 1 cm depth. Plants were treated with different NaF concentrations of 25, 50, 75, 100 mg kg^{-1}in different bed sites which were further treated with *P. fluorescens*.The same set of experiments conducted with EDTA (20 mMkg^{-1} soil). The range of investigated NaF concentration added to the soil was screened with respect to the normal contents of F in soil.

The *P. fluorescens* was applied in a concentration of 10^8 CFU mL^{-1}kg^{-1} per plot. In case of 12 plots, 12,000 ml of *P. fluorescens* in a field area (108 m) added by spraying on soil surfaces after 10 days of germination (Marques *et al*., 2010), whereas the control plots [12 plots with F concentration (added

12,000 ml of NaF) and 3 plots without F and P.F] were supplied with 1000 ml of Millipore water (three replications, Figure 3.2).The field study was conducted for 120 days. The F uptake efficiency of the plant *P. juliflora* was detected at the field scale through the method of Niu *et al.* (2007). Growth parameters and plant biomass were observed at 120 days. The wetland consisted of 27 plots, which included eight F treatment plots along with control and all plots were in triplicates.

Figure 3.1: Overview of plot layout of field experimentat Krashi Vigyan Kendra (KVK) Banasthali, Tonk, Rajasthan, India. T0-Control; T1-25 mg kg^{-1} NaF; T2-50 mg kg^{-1} NaF, T3-75 mg kg^{-1} NaF; T4-100 mg kg^{-1} NaF; T5-25 mg kg^{-1} NaF+P.F; T6-50 mg kg^{-1} NaF+P.F; T7-75 mg kg^{-1} NaF+P.F; T8-100 mg kg^{-1} NaF+P.F. Same set of experiments conducted with EDTA (20 $mMkg^{-1}$ soil).

3.2.17.2 Organ wise accumulation

The bioaccumulation (shoot/soil) and translocation (shoot/root)factor were calculated through the method of Niu *et al.* (2007). The total F content in plant samples was measured through alkali fusion-ion selective technique(McQuaker and Gurney, 1977). For analysis of F accumulation in shoot samples, the leaves were also included.

3.2.17.3 Antioxidant activity

Plant tissue (1 g) was homogenized with 3 ml of 0.1 M of sodium phosphate ($NaPO_4$) buffer (pH 7) in a pre-chilled mortar and pestle and centrifuged at 10,000 rpm for 15 min at 4°C. The supernatant was then used for the estimation of superoxide dismutase (SOD), peroxidase (POD), and catalase (CAT). The activity was measured spectrophotometrically at 560 nm for SOD (Beauchamp and Fridivich, 1971), CAT activity was measured at 240 nm (Luck et al., 1974) and POD activity was determined at 420 nm (Putter et al., 1974).

3.2.17.4 Mineral contents

Root, shoot and leaves samples of *P. juliflora* were collected and oven-dried at 75 °C. Take 1 g of plant sample was digested in 20 ml of 1:1:2 of tri-acid mixture HNO3:HClO4:HF (v:v:v) till appearance of transparent color. After digestion, obtained suspension was filtered through Whatman no. 1 filter paper and diluted with 25 ml double de-ionized water. The total amount of N was determined through the method described in the 3.2.5 section. Total soluble P and K content were analyzed by using (Olsen's) molybdenum blue method and flame photometer 3.2.6 and 3.2.7 section. Diethylenetriaminepenta acetic acid(DTPA) extractable method was explained for the detection of Fe, Mn, Zn and Cu in plant samples in the 3.2.10 section.

3.2.18 Statistical study

The statistical analysis was subject to Analysis of Variance (ANOVA) followed by LSD (Least Significant Difference) post hoc test at probability level to separate means when ANOVA indicated a significant effect in accordance with the experimental design using SPSS (Software Programme for Social Science) software to quantify and evaluate the source of variation and the correlation between the different value were calculated at 5% level of significance. SE (Standard error) values are depicted in tables as ± and in the graphs as error bars. The graphs were prepared by Origin Pro software and analysis of the map image was carried out using ArcGIS (ver. 10.2) software. The map interpretation was done using the Geographic information system (GIS) approach based method for a correct spatial map decision.

4

Results and Discussion

CONTENTS

This chapter incorporates the results and discussion of current studies on a preliminary investigation of physicochemical parameters of soil samples, biochemical characterization of isolated bacterial strains & identification of bacteria by 16S rDNA gene sequence, root elongation assay for F-tolerant bacteria, the effect of F-tolerant bacteria on plant growth, organ-wise F uptake (Section 4.1). Further, effect of microbial consortium *Pseudomonas fluorescens*(P.F) and *Pseudomonas aeuroginosa* (P.A) on plant growth parameters, chlorophyll contents, antioxidant enzyme activity, accumulation study of F hyperaccumulator plant *P. juliflora* were performed (Section 4.2). Effect of chelating agent Ethylene diamine tetraacetic acid (EDTA) and Citric acid (CA)on plant growth parameters, chlorophyll contents, antioxidant enzyme activity, accumulation study of F hyperaccumulator plant *P. juliflora* were also determined(Section 4.3). Furthermore, mineral contents in plants parts (root, shoot and leaves)after treatment with microbes(P.F and P.A) and chelates (EDTA and CA) under different concentration of F and mineral contents in soil after treatment with microbes (P.F and P.A) and chelates (EDTA and CA) were also (Section 4.4). Then, the effect of *Pseudomonas fluorescens* and Ethylene diamine tetra acetic acid on the biochemical, mineral content and (F) accumulation by*P.juliflora* in a field experiment was conducted (Section 4.5).

4.1 PHYSIO-CHEMICAL PARAMETERS OF SOIL SAMPLES

The preliminary investigation of physicochemical characteristics of the soil are shown in (Table 4.1) and samples sites areas of Braham Mandir,Botanical Garden and Krishi Vigyan Kendra regions of Banasthali University(26° 60'N 75° 54'E) Tonk Rajasthan, India (Figure 4.1). The pH of the field soilwas alkaline in nature (7.40-8.02) and was nutrient- poor as the concentrations of EC, O.C, Cl^-, CO_3^{2-}, HCO_3^{2-}, RSC, N, P, K, S, Ca and Mg obtained were low as 1.04-15.35 dSm^{-1},0.29-0.75%,3.20-16.36 $mgkg^{-1}$,0.38-3.63$mgkg^{-1}$,2.78-16.68 mg kg^{-1},1.27-15.49$mgkg^{-1}$,33.46-67.14 $kgha^{-1}$ 3.57-9.36 $mgkg^{-1}$ 102.00-120.09 kg ha^{-1}, 3.15-9.17 mg kg^{-1}, 1.08-4.15 mg $kg^{-1}$and 1.35-2.84 mg kg^{-1}respectively. The micronutrients were also obtained in low concentrations as Fe -1.14-1.76 mg kg^{-1}, Mn-1.12-1.46 mg kg^{-1}, Zn-1.17-1.67 m gkg^{-1}, Cu-1.08-1.77 mg kg^{-1} and F- 40.87-65.58 mg kg^{1}. All the micro and macro nutrients were reported in lower quantities due to the alkaline nature of the contaminated soil. Soil pH has been directly influenced by metal mobility as it was affected by the solubility of metals and their capacity to form soluble chelates in soils.

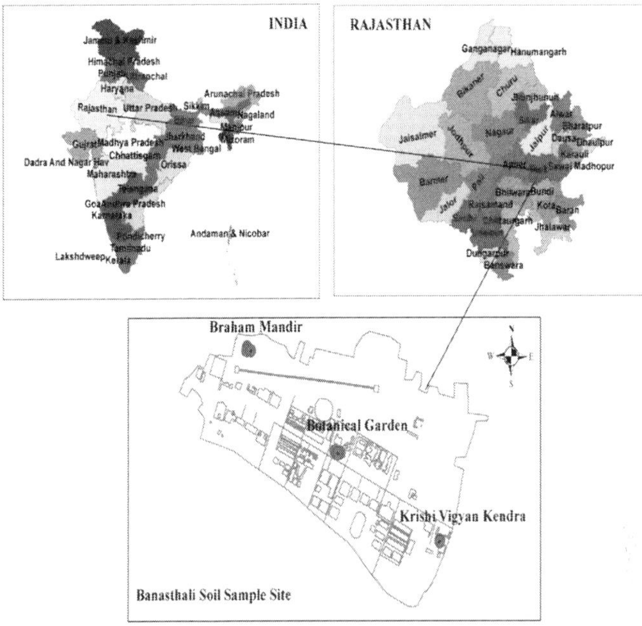

Figure 4.1: Sample collection sites of Banasthali, Tonk, Rajasthan, India as: (BS1-BSIII from Braham Mandir, BSIV-BSV from Botanical Garden, BSVI-BSVII from Krishi Vigyan Kendra).

Isolated *Clostridium sp.*strain was highly resistant to pH and high salinity and could also be a strong candidate for the improvement of plant growth in highly alkaline soils. All soil samples showed a slightly alkaline taste with respect to its pH. Soil samples of the BS1 area showed the lowest electrical conductivity (1.04 dS m^{-1}) whereas the BS7 area was of a higher magnitude of EC values (15.35 dS m^{-1}) on the basis of its mean values. The Ca^{+2} and Mg^{+2} contents (1.08mg kg^{-1}) were found to be lowest in BS1soil samples whereas BS7soil samples contained 4.15mg kg^{-1}(Ca^{+2}+Mg^{+2}) being highest compared to the rest of the soil samples.BS1soilsamples registered the lowest chloride (3.20mg kg^{-1}) and BS7 samples contained the highest (16.36mg kg^{-1}) amount of chloride ions (Table 4.1). Carbonate (0.38mg kg^{-1}) and bicarbonate(2.78mg kg^{-1}) contents were of lowest magnitude in BS1 samples respectively, whereas BS7soil samples registered the highest amount of carbonate and bicarbonate contents being 3.63 and 16.68mg kg^{-1}respectively (Table 4.1). Residual sodium carbonate (RSC) showed a positive correlation with CO_3^{2-}, HCO_3 and (Ca^{+2}+Mg^{+}2) cations in soil. RSC value was highest in the BS7 area (15.49) being lowest (1.27mg kg^{-1}) in BS1 soil samples. Although RSC showed a positive correlation with EC during study, but in Banasthali block RSC of soil was found to be low with less EC. Data pertaining to trace elements in soil, given in (Table 4.1) reveals that iron and manganese contents were comparatively higher than copper and zinc in soil irrespective of area under study. BS7 area soil sample registered a higher amount of iron on the basis of mean values. F contents in soil samples were lowest in BS1(40.87 mg kg^{-1}) and the highest (65.58 mg kg^{-1}) of the same element was recorded in BS7 soil samples.

These twelve parameters were subjected to assess the correlation between the given parameters. Data reveal that standard deviation values of all 12 layers are lower than the mean values, which indicate that the effect of abnormal data on sampling values was not sufficient. Desirable or acceptable limits of these parameters have also been given within the (Table4.1). Covariance matrix of individual quality parameters reveals that layer manganese(-0.115), iron (-0.113), carbonate (-0.135), bicarbonate (-0.483) and residual sodium carbonate (-0.727) were found to be negatively correlated with Ca^{+2} and Mg^{+2} whereas rest of all layers showed positive correlation with presence of Ca^{+2} and Mg^{+2} in soil sample.

Manganese showed a negative correlation (Mn -0.225) with Ca^{+2} and Mg^{+2}, copper (-0.285) and pH (-0.650) and rest all were found to be positively correlated. The iron showed a negative correlation leaving only Mn, Cl, CO_3^{2-} and EC of soil. Leaving aside Ca and Mg, Zn and pH, the Cu had a negative correlation with the other parameters. The presence of Zn with Fe, Cl, HCO_3, pH and RSC showed a negative correlation whereas the rest of all soil parameters were found to be positively correlated with Zn. Chlorine showed a positive correlation with most of the parameters barring Cu (-0.052), Zn (-0.60) and pH (-0.80).

The presence of CO_3 in soil was found to be negatively correlated with Ca and Mg, copper and pH whereas the HCO_3 amount had a negative correlation with Ca, Mg, Mn, Cu, Zn and pH. Further EC was positively correlated with almost all factors barring only Cu whereas RSC showed positive association only with Mn, Cl, CO_3^{2-}, HCO_3, EC and RSC of soil and the rest were highly positively correlated. The correlation matrix of these twelve layers has also been worked out. Data reveals that RSC showed a highly negative correlation among all the layers of soil quality parameters having negative value (-0.552) whereas manganese was highly positively correlated with Chlorine (0.422), Carbonate (0.558)and RSC (0.422) respectively.

Table 4.1: Physicochemical analysis of F-contaminated soil from different sites of Banasthali, Tonk, Rajasthan, India

SP	BS1	BS2	BS3	BS4	BS5	BS6	BS7
SMB	4.40 ±0.17	5.15±0.06	6.86±0.10	6.95±0.27	7.82±0.24	8.96±0.04	15.12±0.43
SMA	2.13±0.45	3.45±0.23	3.89±0.12	4.94±0.34	5.12±0.23	5.76±0.23	6.78±0.34
pH	7.40 ±0.17	8.15±0.06	7.86±0.10	7.55±0.27	7.66±0.24	7.16±0.04	8.02±0.43
E.C	1.04±0.00	2.16±0.02	2.15±0.06	1.10±0.00	1.25±0.00	2.25±0.01	15.35±0.00
O.C	0.29±0.27	0.47±0.04	0.38±0.00	0.50±0.00	0.65±0.00	0.45±0.04	0.75±0.00
Cl^-	3.20±0.56	4.78±0.67	5.29±0.04	8.50±0.20	6.29±0.10	12.78±2.78	16.36±1.56
CO_3^{2-}	0.38±0.28	1.56±0.89	1.26±0.09	2.56±0.10	1.98±0.37	2.89±0.20	3.63±0.90
HCO_3^-	2.78±0.67	3.30±0.36	4.78±0.83	6.95±1.08	6.24±0.89	9.67±0.34	16.68±0.47
RSC	1.27±0.90	2.78±0.49	4.78±1.90	5.98±0.35	8.20±0.48	13.78±0.38	15.49±1.78
N	33.46±0.15	41.48±0.34	50.23±0.00	46.30±0.30	24.41±2.00	43.15±0.80	67.14±0.80
P	3.57±0.24	5.80±0.41	5.64±0.40	5.30±0.20	5.64±0.70	5.70±0.52	9.36±0.60
K	102.00± 2.00	116.00± 2.04	111.00± 0.50	113.08± 1.10	112.02± 1.10	106.01± 1.10	120.09± 1.50
S	3.15±0.00	5.35±0.05	8.36±0.00	8.00±0.00	4.22±0.80	6.60±0.61	9.17±0.00
Ca	1.08±2.90	2.73±0.41	1.50±0.10	1.70±0.60	1.60±0.02	1.55±0.33	4.15±0.50
Mg	1.35±1.89	4.39±1.23	1.60±1.40	1.60±0.90	1.35±0.30	2.86±0.46	2.84±0.40
Fe	1.14±0.00	1.60±0.07	1.70±0.00	1.10±0.00	1.69±0.00	1.26±0.00	1.76±0.00
Mn	1.12±0.03	1.44±0.06	1.20±0.47	1.40±0.67	1.90±0.05	1.84±0.67	1.46±0.60
Zn	1.17±0.00	1.60±0.04	1.26±0.00	1.50±0.00	1.30±0.00	1.55±0.00	1.67±0.00
Cu	1.08±0.00	1.55±0.09	1.78±0.00	1.34±0.00	1.22±0.00	1.27±0.00	1.77±0.00
F	40.87±1.00	42.00±3.03	42.04±0.00	46.80±0.00	52.60±2.00	58.60±2.00	65.58±3.00

Note: Values are mean of three replicates and ± SD (Electrical conductivity-E.C dSm^{-1}),

(Soil moisture before, after- SMB, SMA %, Organic Carbon-O.C %,

Chloride-Cl^-, Carbonate-CO_3^{2-}, Biocarbonate-HCO_3^{2-}, Residual sodium carbonate- RSC,

Sodium-N, Calcium-Ca, Magnesium-Mg, Sulfur-S, Phosphorus-P, Potassium-K , Iron-Fe,

Manganese-Mn, Zinc-Zn, Copper-Cu, Fluoride-F mg kg^{-1}),

SP- Soil parameters (BS1 to BS7)-Banasthali sampling sites.

The $CO_3^{2-}$and HCO_3also showed a strong correlation with the RSC layer. The presence of Zn and Fe did not show any strong correlation with any of the layers under study. Ca^{+2} and Mg^{+2} in soil were found to be highly and negatively correlated with RSC showing the presence of $CO_3^{2-}$and HCO_3thereby higher pH in the soil. Chloride showed a highly positive correlation with Ca^{+2} and Mg^{+2} (Figure 4.2 A, B). Many saline sodic soils contain soluble carbonates besides the excess of soluble salts (Goovaerts, 1999).

Such soils, the topography, soil depth and amount of rainfall might have caused an increase in sodicity and salinity of groundwater. Singh *et al.*(2008) also reported higher pH and EC of soil due to low recharging of wells and soils leaving soluble salts accumulated in the subsurface layer. Higher EC in the soil might be due to the fact that salts reach quickly in groundwater in shallow soils.

The declining water table in certain parts of Rajasthan and Gujarat has also been reported to being responsible for fluctuations in sodicity and salinity of soil (Samara, 2002).

F has low mobility in soil as compared to other metals and does not accumulate in upper soil horizons, but in slightly acidic soils, it is more soluble and greater leaching. Most of the F in the soil is insoluble but soluble F is important to the plants.

The higher concentration of F levels in the soil at low pH which increases mobility to uptake via the root and is more available to the plants. In India,

GIS has been introduced in various fields like optimizing land use plans, characterization of soil quality, waste's land and management of salt -affected soils.

Soil is an essential parameter to be studied for the sustainable development of agriculture and human life.

The advent of information technology has developed tools like GPS and GIS which help in the spatial characterization of soil.

The maps generated through GPS and GIS delineate homogenous units to decide on the size and collecting a systematic set geo-referenced samples and generating spatial data about soil (Sood *et al.*, 2004; Sarkar, 2011).

The soil quality parameters were used to carve out spatial distribution maps of all soil quality parameters of the present study. Proper legends have also been developed over each spatial map to describe the relative distribution of each quality parameter of soil property.

The soil nutrients contents are also susceptible to F concentration and it should be considered as F contaminated soil.

From the present study, it can be concluded that the use of remote sensing and GIS for studying potential F contaminated soil which generate basic ideas of the soil contamination with F. Also, a GIS -based study can be very useful for decision maker's exploration of groundwater to understand the potential of present research work on fluoride contaminated soil.

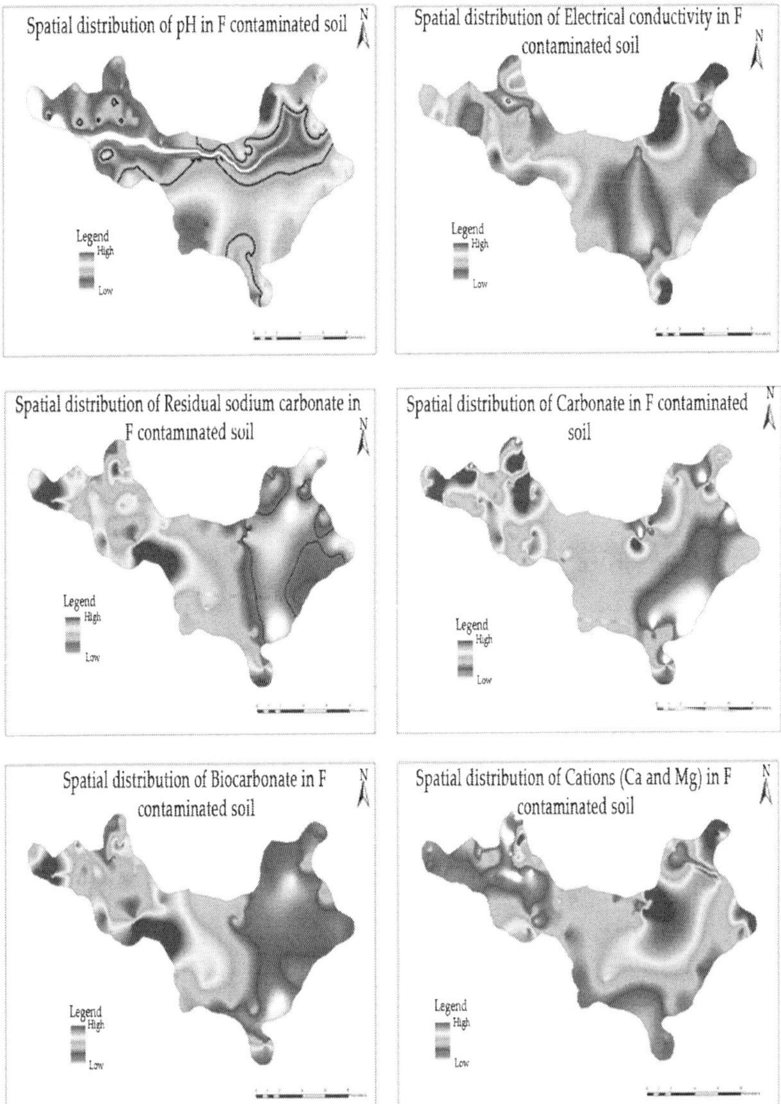

Figure 4.2.A: Spatial distribution in pH, EC, Residual sodium carbonate, Carbonate, Biocarbonate and Cation parameters of Banasthali Vidyapith, Rajasthan, India.

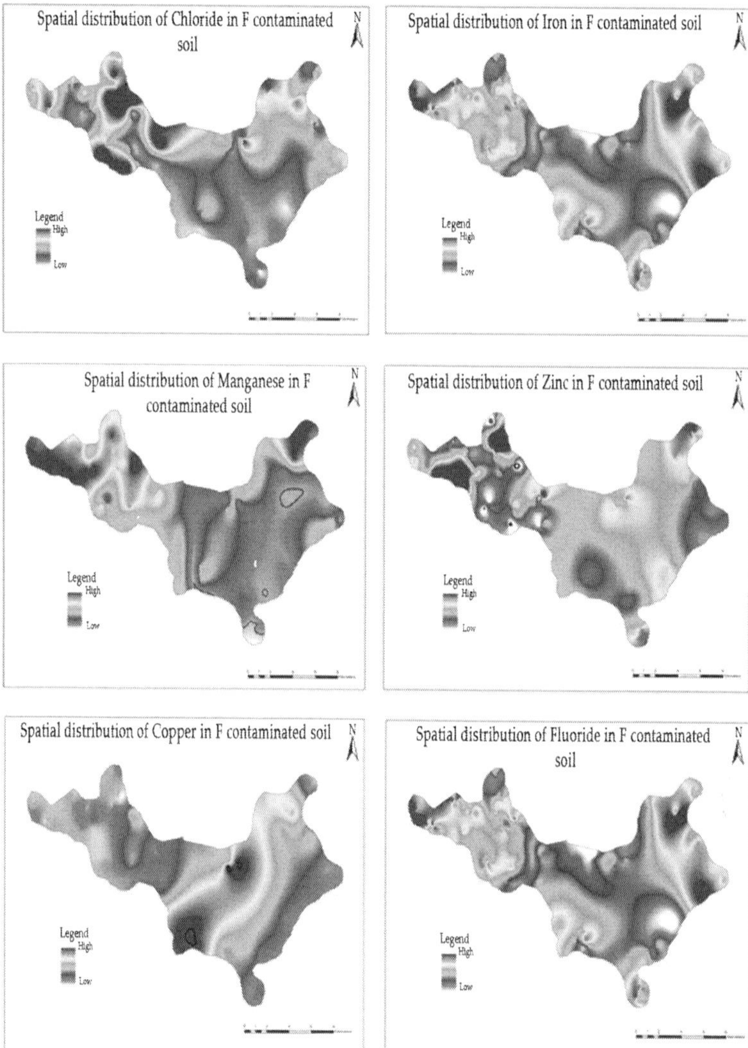

Figure 4.2.B: Spatial distribution in Chloride, Iron, Manganese, Zinc, Copper and Fluoride of Banasthali Vidyapith, Rajasthan, India.

4.2 BIOCHEMICAL CHARACTERIZATION OF THE ISOLATED BACTERIAL STRAINS

The two bacteria were isolated from twelve strains and selected for further characterization. The two isolated bacterial strains were identified as *Clostridium sp.* It was observed that the colonies of isolated bacteria were mostly round in structure and cream in color (Figure 4.3. A, B). Both of the bacterial strains were gram positive. Further biochemical studies were performed to identify them correctly (Table 4.2and Table 4.3). The antibiotic susceptibility of bacteria to different antibiotics (ampicillin, streptomycin, gentamycin, tetracycline and chloramphenicol) showed a zone of inhibition around the bacteria colony. All bacterial strains that were selected showed positive results towards catalase activity, amino acid deaminase test, ammonia production, nitrate reductase test, citrate test, MR-VP indole test as per the standard methods. Total of 35 tests were performed for the identification of carbohydrate fermentation test but only sixteen gave positive result (Table 4.3) which were (Sorbitol, Fructose, Inositol, Salicin, Glycerol, Sodium gluconate, Inulin, L-Arabinose, Sucrose, Trehalose, Dextrose, Maltose, ONPG, Adonitol, a-Methyl-D-glucoside and Erythritol) produces yellow color.

4.3 IDENTIFICATION OF ISOLATE BACTERIA BY 16S RDNA GENE

The two isolated bacterial strains were genomically characterized by 16S rDNA gene sequencing (Figure s 4.4 and 4.5). The BLAST alignment search tool of NCBI Genbank database were performed on 16S rDNA gene sequences.

The isolated strains showed a relatively high degree of diversity (99%) on the basis of phylogenctic analysis (Gene Bank accession no.1772642-KP136288, 1774032- KP136289, 1774035- KP136290, 1774037-KP136291).

The 16S rDNA gene data showed that the selected strains were related to several differently named species. On the basis of 16S rDNA gene sequencing and molecular phylogeny analysis, these strains were described as *Clostridium sp*.

Phenotypic analyses, as well as 16S rDNA gene-based analyses, showed a high degree of diversity reported in the present study for fluoride tolerant bacterial strains.

The study concerned about 16 nucleotide sequences and codon positions incorporated were 1st+2nd+3rd+Noncoding. All the positions containing gaps and missing data were eliminated. There were a total of 1224 and 1259 positions in the final dataset (Kimura, 2004 and Kim *et al*., 2012).

Table 4.2: Biochemical tests for F-tolerant bacteria (FTB) from the Banasthali region of Tonk, Rajasthan, India

Characteristics					
Test	**FTB1**	**FTB2**	**Test**	**FTB1**	**FTB2**
Gram staining	+	+	Catalase test	+	+
Motility	+	+	Oxidase test	+	+
Hydrogen sulfide production test	+	+	Ammonia production test	+	+
Indole acetic acid test	+	+	Sugar fermentation test	+	+
Methyl red Test	+	+	Nitrate reduction test	+	+
Voges Proskauer test	+	+	Hydrogen cyanide Production Tesr	+	+
Antibiotic Test	+	+	Carbohydrate utilization test		

Note: + = positive, - = negative;

FTB1 and FTB2

(Fluoride tolerant bacteria = Clostridium sp).

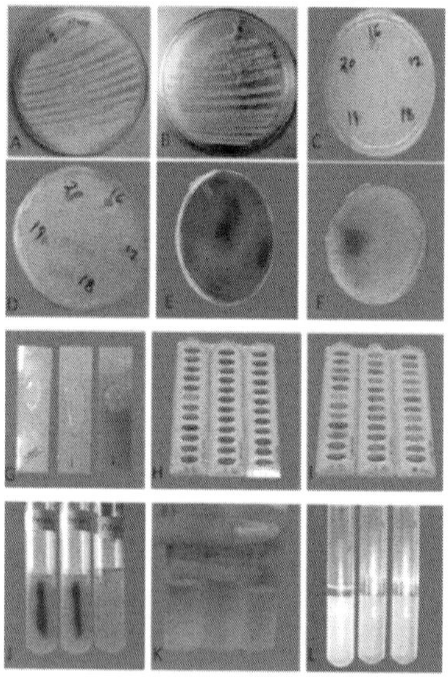

Figure 4.3.A: Isolated F-tolerant bacteria (A and B) from Banasthali region of Tonk, Rajasthan, India and biochemical characterization for selected F tolerant bacteria (C-D) Antibiotic sensitivity test, (E-F) Ammonia production, (G) Catalase, (H-I) Carbohydrate utilization test, (J) Motility and hydrogen sulfide production test, (K) Sugar fermentation test (L) Nitrate reduction test.

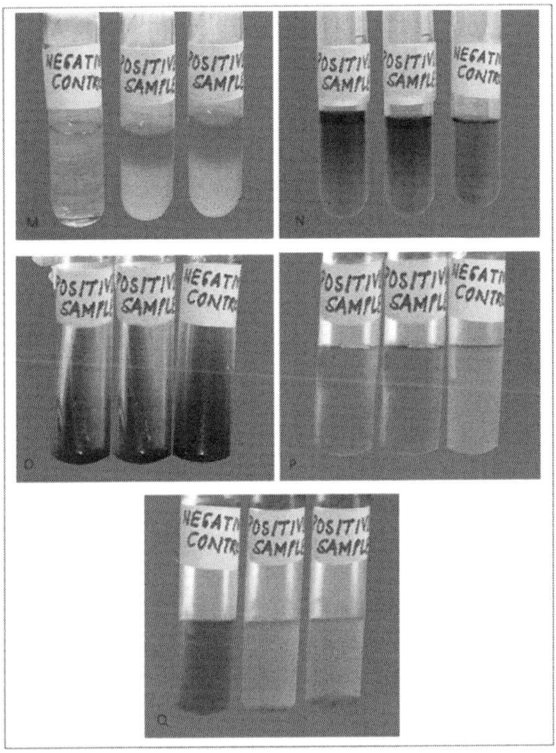

Figure 4.3.B: Biochemical characterization for selected F tolerant bacteria (M) Indole acetic acid test (N) Hydrogen cyanide test (O) Oxidase test, (P) Methyl red test and (Q) Voges Proskauer test.

Table 4.3: Carbohydrate utilization tests for F-tolerant bacteria (FTB) from the Banasthali region of Tonk Rajasthan, India

S.No.	CT	FTB1	FTB2	CT	FTB1	FTB2
1.	Lactose	-	-	Inositol	+	-
2.	Xylose	-	+	Sorbitol	+	-
3.	Maltose	+	+	Mannitol	-	-
4.	Fructose	+	+	Adonitol	+	-
5.	Dextrose	+	+	Arabitol	-	-
6.	Galactose	-	-	Erytritol	+	-
7.	Raffinose	-	-	a-Methyl-D-glucoside	+	-
8.	Trehalose	+	+	Rhamnose	-	-
9.	Melibiose	-	+	Cellobiose	-	-

10.	Sucrose	+	+	Melezitose	-	+
11.	L-Arabi-ose	+	-	a-Methyl-Dmanno-side	-	-
12.	Mannose	-	-	Xylitol	-	-
13.	Inulin	+	+	ONPG	+	-
14.	Sodium gluconate	+	+	Esculin hydrolysis	-	-
15.	Glycerol	+	+	D-Arabinose	-	+
16.	Salicin	+	+	Citrate utilization	-	-
17.	Dulcitol	-	-	Malonate utilization	-	-
18.	Sobose	-	+	Control	-	-

Note: + = positive, - = negative; FTB1 and FTB2 (Fluoride tolerant bacteria = Clostridium sp).

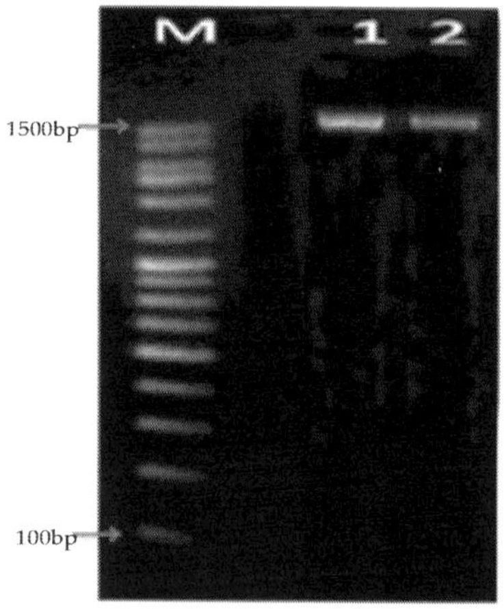

Figure 4.4: 16S rDNA gene amplicon of isolated F tolerant bacteria.

4.3.1 Chromatogram Data File Detail (FTB1): The following chromatogram files will be attached separately along with this report:

1. PSB(I)_704F_S011876_C01_011.ab1: Data obtained with Forward primer
2. PSB(I)_907R_S011876_D01_009.ab1: Data obtained with Reverse primer

FTB(I)_704F_S011876_C01_011 (698 bp)

CTACTGGGCTTTAACTGACGCTGAGGCACGAAAGCATGGGTAG-
CAAACAGGATTAGATACCCTGGTAGTCCATGCCGTAAACGAT-
GATTACTAGGTGTGGGGGTCTGACCCCTTCCGTGCCGGAGT-
TAACACAATAAGTAATCCACCTGGGGAGTACGGCCGCAAGGTT-
GAAACTCAAAGGAATTGACGGGGGCCCGCACAAGCAGTG-
GAGTATGTGGTTTAATTCGAAGCAACGCGAAGAACCTTAC-
CAGGTCTTGACATCCAACTAACGAGGCAGAGATGCATTAGGT-
GCCCTTCGGGGAAAGTTGAGACAGGTGGTGCATGGTTGTC-
GTCAGCTCGTGTCGTGAGATGTTGGGTTAAGTCCCGCAAC-
GAGCGCAACCCTTGTGATTAGTTGCTACGCAAGAGCACTCTA-
ATCAGACTGCCGTTGACAAAACGGAGGAAGGTGGGGACGAC-
GTCAAATCATCATGCCCCTTATGACCTGGGCTACACACGTAC-
TACAATGGTCGCCAACAGAGGGAAGCCAAGCCGCGAGGTG-
GAGCAAACCCCCAAAAGCGATCTCAGTTCGGATTGTAGGCTG-
CAACCCGCCTACATGAAGTTGGAATTGCTAGTAATCGCGGAT-
CAGCATGCCGCGGTGAATACGTTCCCGGGCCTTGTACACACC-
GCCCGTCACACCATGGGAGCCGGTAATACGCGAAGC

FTB(I)_907R_S011876_D01_009 (696 bp)

ACGGCATGGACTACCAGGGTATCTAATCCTGTTTGCTACCCAT-
GCTTTCGTGCCTCAGCGTCAGTTAAAGCCCAGTAGGCC-
GCCTTCGCCACTGGTGTTCCTCCCGATCTCTACGCATTTCACC-
GCTACACCGGGAATTCCGCCTACCTCTACTTCACTCAAGCCTTC-
CAGTTTCGAACGCAATTTGTGGGTTAAGCCCACAGCTTTCAC-
GCCCGACTTAAAAAGCCGCCTACGCACCCTTTACACCCAG-
TAAATCCGGACAACGCTTGCTCCCTACGTATTACCGCGGCT-
GCTGGCACGTAGTTAGCCGGAGCTTCCTCCTTGGCTACCGT-
CATTATCTTCACCAAGGACAGAGGTTTACAATCCGAAGACC-
GTCTTCCCTCACGCGGCATTGCTGCATCAGAGTTTCCTCCATT-
GTGCAATATCCCCCACTGCTGCCTCCCGTAGGAGTCTGGGC-
CGTGTCTCAGTCCCAATGTGGCCGTTCAGTCTCTCAACCCG-
GCTACCGATCGTCGCTTTGGTGGGCCTTTACCCCGCCAACTG-
GCTAATCGGACGCGAGTCCATCTTTCAGCGGATTGCTCCTTT-
GATACCGGGGCCATGCGACCCCAATATATTATGCGGTATTAGC-
GTTCTTTTCAGAACGTTATCCCCCTCTGAAAGGCAGGTTACT-
CACGCGTTACTCACCCGTCCGCCACTAAG

4.3.1.1 Consensus sequence FTB(I) (1316 bp)

CTTAGTGGCGGACGGGTGAGTAACGCGTGAGTAACCT-
GCCTTTCAGAGGGGGATAACGTTCTGAAAAGAACGCTAATAC-
CGCATAATATATTGGGGTCGCATGGCCCCGGTATCAAAGGAG-
CAATCCGCTGAAAGATGGACTCGCGTCCGATTAGCCAGTTG-
GCGGGGTAAAGGCCCACCAAAGCGACGATCGGTAGCC-
GGGTTGAGAGACTGAACGGCCACATTGGGACTGAGACAC-
GGCCCAGACTCCTACGGGAGGCAGCAGTGGGGGATATT-
GCACAATGGAGGAAACTCTGATGCAGCAATGCCGCGT-
GAGGGAAGACGGTCTTCGGATTGTAAACCTCTGTCCTTG-
GTGAAGATAATGACGGTAGCCAAGGAGGAAGCTCCGGCTA-
ACTACGTGCCAGCAGCCGCGGTAATACGTAGGGAGCAAGC-
GTTGTCCGGATTTACTGGGTGTAAAGGGTGCGTAGGCG-
GCTTTTTAAGTCGGGCGTGAAAGCTGTGGGCTTAACCCA-
CAAATTGCGTTCGAAACTGGAAGGCTTGAGTGAAGTAGAG-
GTAGGCGGAATTCCCGGTGTAGCGGTGAAATGCGTAGAGA-
TCGGGAGGAACACCAGTGGCGAAGGCGGCCTACTGGGCTT-
TAACTGACGCTGAGGCACGAAAGCATGGGTAGCAAACAG-
GATTAGATACCCTGGTAGTCCATGCCGTAAACGATGATTAC-
TAGGTGTGGGGGGTCTGACCCCTTCCGTGCCGGAGTTAA-
CACAATAAGTAATCCACCTGGGGAGTACGGCCGCAAGGTT-
GAAACTCAAAGGAATTGACGGGGGCCCGCACAAGCAGTG-
GAGTATGTGGTTTAATTCGAAGCAACGCGAAGAACCTTAC-
CAGGTCTTGACATCCAACTAACGAGGCAGAGATGCATTAGGT-
GCCCTTCGGGGAAAGTTGAGACAGGTGGTGCATGGTTGTCGT-
CAGCTCGTGTCGTGAGATGTTGGGTTAAGTCCCGCAACGAGC-
GCAACCCTTGTGATTAGTTGCTACGCAAGAGCACTCTAAT-
CAGACTGCCGTTGACAAAACGGAGGAAGGTGGGGACGACGT-
CAAATCATCATGCCCCTTATGACCTGGGCTACACACGTACTA-
CAATGGTCGCCAACAGAGGGAAGCCAAGCCGCGAGGTGGAG-
CAAACCCCCAAAAGCGATCTCAGTTCGGATTGTAGGCTGCAAC-
CCGCCTACATGAAGTTGGAATTGCTAGTAATCGCGGATCAGCAT-
GCCGCGGTGAATACGTTCCCGGGCCTTGTACACACCGCCCGTCA-
CACCATGGGAGCCGGTAATACGCGAAGC

4.3.1.2 Sequence producing significant alignments

Accession	Description	Max score	Total score	Query cover-age	E val-ue	Max ident
HG726040.1	Clostridium sp.n isolate JCD	2425	2425	100%	0.0	99%
JN091083.1	Clostridium sp. MSTE9	2398	2398	100%	0.0	99%
JN873208.1	Uncultured Clostridiales bacterium clone JXS2-6	2394	2394	100%	0.0	99%
JQ993475.1	Uncultured Clostridia bacte-rium clone HsjTcB_106	2386	2386	100%	0.0	99%
JQ993476.1	Uncultured Clostridia bacte-rium clone HsjTcB_116	2386	2386	100%	0.0	99%
M59116.1	Clostridium sporosphaeroides	2353	2353	100%	0.0	99%
HQ789326.1	Uncultured organism clone ELU0116-T290-S-NI_000107	2326	2326	96%	0.0	99%
EF434352.1	Uncultured Clostridiales bacte-rium clone 1099982248072	2324	2324	96%	0.0	99%
HM306081.1	Uncultured bacterium clone ncd860f03c1	2318	2318	96%	0.0	99%
GQ011067.1	Uncultured bacterium clone nbw790e11c1	2318	2318	96%	0.0	99%
HQ772705.1	Uncultured organism clone ELU0071-T471-S-NIPCRAM-gANa_000249	1989	1989	99%	0.0	94%
HQ772657.1	Uncultured organism clone ELU0071-T471-S-NIPCRAM gANa_000201	1989	1989	99%	0.0	94%
HQ770081.1	Uncultured organism clone ELU0065-T439-S-NIPCRAM-gANa_000615	1989	1989	100%	0.0	94%
EF405280.1	Uncultured bacterium clone SJTU_G_03_48	1989	1989	99%	0.0	94%
AY684393.3	Uncultured bacterium clone HuAC05	1989	1989	99%	0.0	94%

4.3.2 Chromatogram Data File Detail (FTB2): The following chromatogram files will be attached separately along with this report:

1. PSB(II)_704F_S011876_E01_007.ab1: Data obtained with Forward primer

2. PSB(II)_907R_S011876_F01_005.ab1: Data obtained with Reverse primer

FTB(II)_704F_S011876_E01_007 (649 bp)

TGGTAGTCCATGCCGTAAACGATGATTACTAGGT-
GTGGGGGTCTGACCCCTTCCGTGCCGGAGTTAACACAATA-
AGTAATCCACCTGGGGAGTACGGCCGCAAGGTTGAAACT-
CAAAGGAATTGACGGGGGCCCGCACAAGCAGTGGAGTAT-
GTGGTTTAATTCGAAGCAACGCGAAGAACCTTACCAGGTCTT-
GACATCCAACTAACGAGGCAGAGATGCATTAGGTGCCCTTC-
GGGGAAAGTTGAGACAGGTGGTGCATGGTTGTCGTCAGCTCGT-
GTCGTGAGATGTTGGGTTAAGTCCCGCAACGAGCGCAACCCTT-
GTGATTAGTTGCTACGCAAGAGCACTCTAATCAGACTGCCGTT-
GACAAAACGGAGGAAGGTGGGGACGACGTCAAATCATCAT-
GCCCCTTATGACCTGGGCTACACACGTACTACAATGGTCGCTAAC-
AGAGGGAAGCCAAGCCGCGAGGTGGAGCAAACCCCCAAAAGC-
GATCTCAGTTCGGATTGTAGGCTGCAACCCGCCTACAT-
GAAGTTGGAATTGCTAGTAATCGCGGATCAGCATGCCGCGGGT-
GAATACGTTCCCGGGCCTTGTACACACCGCCCGTCACAC-
CATGGGAGCCGGTAATACCCGAAGCCAATAGTCAAAC

FTB(II)_907R_S011876_F01_005 (745 bp)

CATGGACTACCAGGGTATCTAATCCTGTTTGCTACCCATGCTTTC-
GTGCCTCAGCGTCAGTTAAAGCCCAGTAGGCCGCCTTCGC-
CACTGGTGTTCCTCCCGATCTCTACGCATTTCACCGCTA-
CACCGGGAATTCCGCCTACCTCTACTTCACTCAAGCCTTC-
CAGTTTCGAACGCAATTTGTGGGTTAAGCCCACAGCTTTCAC-
GCCCGACTTAAAAAGCCGCCTACGCACCCTTTACACCCAG-
TAAATCCGGACAACGCTTGCTCCCTACGTATTACCGCGGCT-
GCTGGCACGTAGTTAGCCGGAGCTTCCTCCTTGGCTACCGT-
CATTATCTTCACCAAGGACAGAGGTTTACAATCCGAAGACC-
GTCTTCCCTCACGCGGCATTGCTGCATCAGAGTTTCCTCCATT-
GTGCAATATCCCCCACTGCTGCCTCCCGTAGGAGTCTGGGC-
CGTGTCTCAGTCCCAATGTGGCCGTTCAGTCTCTCAACCCG-
GCTACCGATCGTCGCTTTGGTGGGCCTTTACCCCGCCAACTG-
GCTAATCGGACGCGAGTCCATCTTTCAGCGGATTGCTCCTTT-
GATACCAAGGCCATGCGACCCCAATATATCATGCGGTATTAGC-
GTTCTTTTCAGAACGTTATCCCCCTCTGAAAGGCAGGTTACT-
CACGCGTTACTCACCCGTCCGCCACTAAGTAAGAGCTAAG-

CAAGCTTCTCTCTTACTCCGTTCGACTTGCATGTGTTAGGCA

4.3.2.1 Consensus sequence FTB(II) (1382 bp)

TGCCTAACACATGCAAGTCGAACGGAGTAAGAGAGA-
AGCTTGCTTAGCTCTTACTTAGTGGCGGACGGGTGAGTA-
ACGCGTGAGTAACCTGCCTTTCAGAGGGGGATAACGTTCT-
GAAAAGAACGCTAATACCGCATGATATATTGGGTCGCATG-
GCCTTGGTATCAAAGGAGCAATCCGCTGAAAGATGGACTCGC-
GTCCGATTAGCCAGTTGGCGGGGTAAAGGCCCACCAAAGC-
GACGATCGGTAGCCGGGTTGAGAGACTGAACGGCCA-
CATTGGGACTGAGACACGGCCCAGACTCCTACGGGAGGCAG-
CAGTGGGGGATATTGCACAATGGAGGAAACTCTGATGCAG-
CAATGCCGCGTGAGGGAAGACGGTCTTCGGATTGTAAACCTCT-
GTCCTTGGTGAAGATAATGACGGTAGCCAAGGAGGAAGCTC-
CGGCTAACTACGTGCCAGCAGCCGCGGTAATACGTAGGGAG-
CAAGCGTTGTCCGGATTTACTGGGTGTAAAGGGTGCGTAG-
GCGGCTTTTTAAGTCGGGCGTGAAAGCTGTGGCTTAACCCA-
CAAATTGCGTTCGAAACTGGAAGGCTTGAGTGAAGTAGAG-
GTAGGCGGAATTCCCGGTGTAGCGGTGAAATGCGTAGAGA-
TCGGGAGGAACACCAGTGGCGAAGGCGGCCTACTGGGCTT-
TAACTGACGCTGAGGCACGAAAGCATGGGTAGCAAACAG-
GATTAGATACCCTGGTAGTCCATGCCGTAAACGATGATTAC-
TAGGTGTGGGGGGTCTGACCCCTTCCGTGCCGGAGTTAA-
CACAATAAGTAATCCACCTGGGGAGTACGGCCGCAAGGTT-
GAAACTCAAAGGAATTGACGGGGGCCCGCACAAGCAGTG-
GAGTATGTGGTTTAATTCGAAGCAACGCGAAGAACCTTAC-
CAGGTCTTGACATCCAACTAACGAGGCAGAGATGCATTAGGT-
GCCCTTCGGGGAAAGTTGAGACAGGTGGTGCATGGTTGTC-
GTCAGCTCGTGTCGTGAGATGTTGGGTTAAGTCCCGCAAC-
GAGCGCAACCCTTGTGATTAGTTGCTACGCAAGAGCACTCTA-
ATCAGACTGCCGTTGACAAAACGGAGGAAGGTGGGGACGAC-
GTCAAATCATCATGCCCCTTATGACCTGGGCTACACACGTAC-
TACAATGGTCGCTAACAGAGGGAAGCCAAGCCGCGAGGTG-
GAGCAAACCCCCAAAAGCGATCTCAGTTCGGATTGTAGGCTG-
CAACCCGCCTACATGAAGTTGGAATTGCTAGTAATCGCGGAT-
CAGCATGCCGCGGGTGAATACGTTCCCGGGCCTTGTACACACC-
GCCCGTCACACCATGGGAGCCGGTAATACCCGAAGCCAATAGT-
CAAAC.

4.3.2.2 Sequence producing significant alignments

Accession	Description	Max score	Total score	Query coverage	E value	Max ident
JN091083.1	Clostridium sp. MSTE9	2547	2547	100%	0.0	99%
JN873208.1	Uncultured Clostridiales bacterium clone JXS2-6	2531	2531	100%	0.0	99%
HQ789326.1	Uncultured organism clone ELU0116-T290-S-NI_000107	2451	2451	100%	0.0	99%
EF434352.1	Uncultured Clostridiales bacterium clone 1099982248072	2438	2438	95%	0.0	99%
HG726040.1	Clostridium sp. JCD partial	2420	2420	96%	0.0	99%
JQ993475.1	Uncultured Clostridia bacterium clone HsjTcB_106	2385	2385	96%	0.0	99%
JQ993476.1	Uncultured Clostridia bacterium clone HsjTcB_116	2385	2385	96%	0.0	99%
M59116.1	Clostridium sporosphaeroides	2351	2351	96%	0.0	99%
HM306081.1	Uncultured bacterium clone ncd860f03c1	2294	2294	92%	0.0	99%
GQ011067.1	Uncultured bacterium clone nbw790e11c1	2294	2294	92%	0.0	99%
EF401710.1	Uncultured bacterium clone SJTU_D_14_21	2026	2026	100%	0.0	93%
HQ812133.1	Uncultured organism clone ELU0167-T400-S-NIPCRAMgA-Na_000442	2025	2025	99%	0.0	93%
HQ746235.1	Uncultured organism clone ELU0020-T93-S-NI_000381	2025	2025	100%	0.0	93%
GQ898095.1	Uncultured bacterium clone S1-1	2025	2025	99%	0.0	93%
GQ896665.1	Uncultured bacterium clone C1-123	2025	2025	99%	0.0	93%

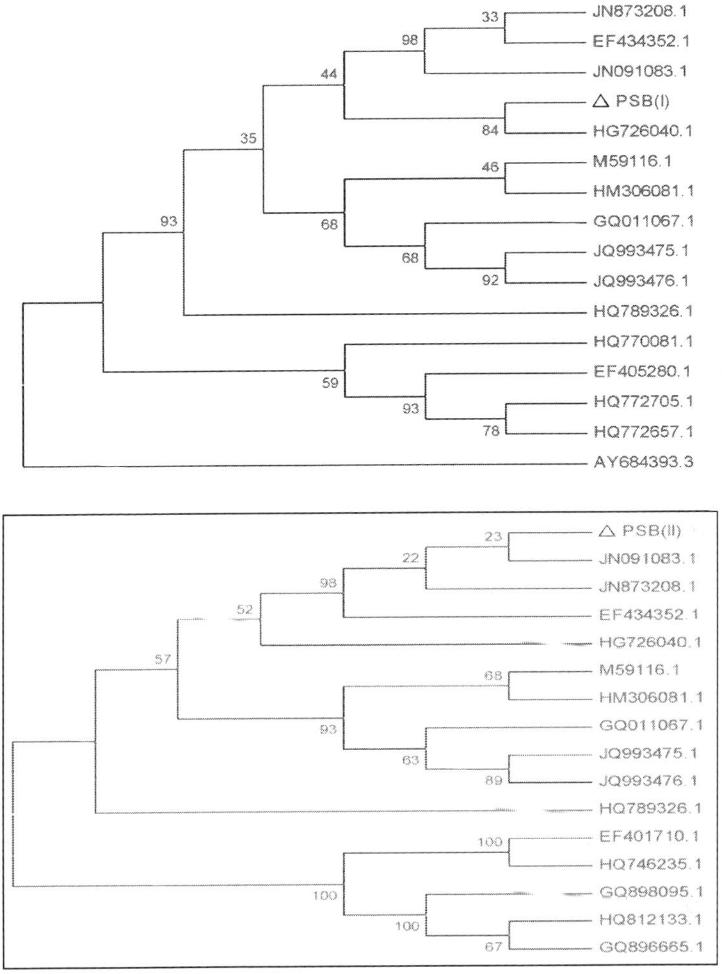

Figure 4.5: Phylogenetic analysis of isolated F-tolerant bacteria (FTB1 and FTB2) based on the nucleotide sequences of 16S rDNA gene.

4.4 ROOT ELONGATION ASSAY FOR F- TOLERANT BACTERIA

Inoculated treated seeds with different concentrations of F were placed on filter paper on the plate. The root length of three replicate was calculated which are shown in (Figure 4.6). Both organisms were *Clostridium sp.* that helped the seedlings to grow and increased the root length.

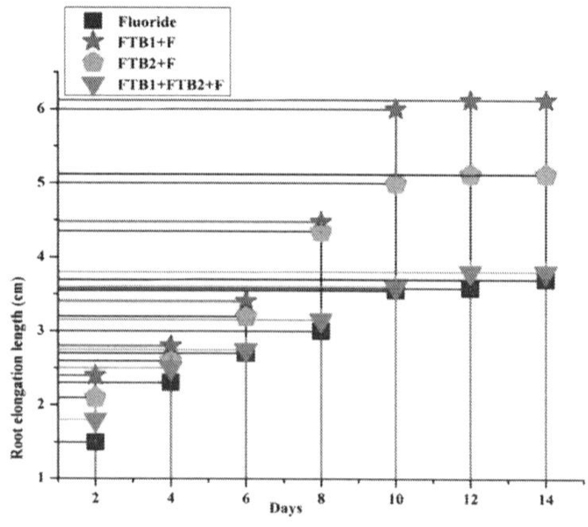

Figure 4.6: Root elongation assay for F-tolerant bacteria(FTB1, FTB2and FTB1+FTB2).

(1)

(2)

Plate 4.4.1.A: Effect of FTB1 and FTB2 on plant growth for 120 days of pot experiment with different concentrations of NaF viz. control, 25, 50, 75 and 100 mg kg^{-1} soil.

(1) FTB1 (Fluoride tolerant bacteria 1)

(2) FTB2 (Fluoride tolerant bacteria 2)

4.5 EFFECT OF F-TOLERANT BACTERIA ON PLANT GROWTH

The effect of *Clostridium sp.*(FTB1, FTB2 and FTB1+FTB2) on plant growth is shown in (Figure 4.7 and Table 4.4). The plant showed promoting effect after adding FTB1. Even though the length and biomass of plant tissues were significantly ($p<0.05$) affected at 100 mg kg-1 NaF with inoculation, *P. juliflora* did not show any toxic symptom under F stress. The inoculation of *Clostridium sp.*(FTB1, FTB2 and FTB1+FTB2)with F stress significantly ($p\leq0.05$) increased both the shoot length (14.84 cm) and the shoot biomass (FW 0.356 g and DW 0.316 g) after adding FTB1 as compared to control group. Meanwhile, FTB1 promoted root length (23.03 cm) and root biomass production (FW 0.284 g, and DW 0.225 g) of plants under F stress condition as compared to FTB2 and combined FTB1+FTB2. The FTB1 and FTB2 had helped inF accumulation with increasing plant biomass. The root length promoting effect of the selected strains has been due to their ability to act as a sink for ACC. The ACC deaminase activity is important for PGPB bacteria to increase the total root surface area (Dell's Amico *et al.*, 2008; Spaepen and Vanderleyden, 2010). The isolated bacteria such as *Agrobacterium ,Achromobacter, Asorhizobium, Bacillus, Burkholderia, Cytophaga* and *Pseudomonas* were shown promoting effect under stress condition. Some of these can be beneficial to plant growth, directly or indirectly (Glick *et al.*, 1999; Belimov *et al.*, 2005; Li *et al.*, 2007; Ma *et al.*, 2009; Lugtenberg and Kamilova, 2009; Ali *et al.*, 2013; Chaudhary and Khan, 2018).

Table 4.4.1: Root and shoot length of *P. juliflora* plants grown in soil for 30, 60, 90 and 120 days in different F concentration viz. control, 25, 50, 75 and 100 mg kg^{-1} NaF soil after treatment with microbes FTB1 and FTB2

Biomass content								
Root length (cm)					**Shoot length (cm)**			
Treatment	**30 days**	**60 days**	**90 days**	**120 days**	**30 days**	**60 days**	**90 days**	**120 days**
Control	9.74±0.64	23.03±0.58	24.51±1.11	25.69±1.06	7.26±0.28	11.89±0.58	13.40±0.16	16.03±0.89
Control 25	21.18±0.16	18.73±0.05	20.81±0.61	23.36±2.11	5.47±0.15	10.14±0.06	11.85±0.54	14.23±0.27
25+ FTB1	23.07±1.41	28.58±0.74	29.81±0.51	30.11±0.00	8.50±0.28	13.48±1.15	15.66±0.79	19.14±0.12
Control 50	16.29±0.25	17.98±0.58	19.40±0.83	23.37±0.28	5.37±0.25	8.28±0.44	10.96±0.72	12.37±0.33
50+ FTB1	17.37±0.52	19.77±1.29	25.33±0.66	27.83±1.06	7.77±0.90	9.84±0.74	13.29±0.67	16.55±2.70
Control 75	14.18±0.16	15.22±0.58	18.52±0.83	22.13±0.52	4.57±0.07	7.81±0.12	9.51±0.45	11.29±0.17
75 + FTB1	17.48±0.71	18.57±0.19	22.18±1.00	26.69±1.00	7.13±0.11	9.55±0.29	12.14±0.16	15.73±0.77
Control 100	12.11±0.11	14.69±0.47	17.36±0.51	19.28±1.79	4.32±0.37	6.58±0.51	8.77±0.00	10.08±0.07
100 + FTB1	14.64±0.20	17.07±1.58	21.96±0.45	23.03±0.60	6.20±0.06	8.15±0.13	10.18±0.0.06	14.84±0.89
Control 25	15.25±0.35	17.77±0.99	20.99±0.50	22.06±0.55	5.36±0.38	8.32±0.44	10.44±0.48	13.11±0.11
25 + FTB2	18.18±0.16	21.33±2.62	24.85±0.48	27.37±0.53	6.82±0.15	11.55±0.69	14.18±0.16	18.21±0.69
Control 50	14.14±0.12	16.88±0.67	19.15±0.64	21.37±0.16	5.11±0.09	7.25±0.27	9.41±0.34	12.29±0.25
50 + FTB2	15.33±0.11	18.69±0.78	22.29±0.22	25.54±1.05	6.62±0.30	9.78±0.67	12.88±0.48	17.22±0.29
Control 75	13.11±0.11	13.52±0.61	18.26±0.39	20.00±0.29	4.29±0.12	6.21±0.11	8.98±0.01	10.14±0.06

75 + FTB2	14.54±0.22	17.48±0.35	21.03±1.57	24.03±0.50	5.19±0.17	8.94±0.05	11.47±0.41	16.45±0.58
Control 100	10.32±0.39	11.21±1.27	14.44±0.19	19.59±0.42	3.58±0.42	5.15±0.05	7.59±0.52	9.52±0.21
100 + FTB2	12.55±0.28	15.28±0.55	20.77±0.57	21.63±1.58	4.76±0.21	7.80±0.11	9.033±0.45	15.18±0.16
Control 25	16.18±0.31	17.36±1.38	19.47±0.72	21.92±1.54	5.22±0.22	5.99±0.55	7.44±0.84	10.66±1.41
25 + FTB1+FTB2	17.66±0.57	18.48±0.50	20.55±1.67	23.25±0.95	7.59±1.02	7.62±0.65	9.31±0.88	11.55±1.44
Control 50	12.18±0.16	13.99±1.63	16.55±1.11	18.14±0.56	4.40±0.16	5.01±0.47	6.46±0.91	9.92±0.28
50+ FTB1+FTB2	14.22±0.19	15.33±1.19	17.40±0.65	18 84±1.00	5.48±0.50	6.73±0.72	8.25±0.70	10.25±1.87
Control 75	11.18±1.16	12.29±0.06	15.62±1.16	16.81±0.83	4.11±0.11	4.37±0.54	6.40±1.33	8.63±0.67
75+ FTB1+FTB2	12.33±0.29	13.47±1.05	15.51±1.00	17.66±1.11	5.92±0.34	5.66±0.29	7.11±0.96	8.62±0.83
Control 100	9.03±1.76	11.22±1.00	13.99±1.63	15.99±0.55	3.13±0.17	3.81±0.89	5.44±0.73	7.70±0.78
100+ FTB1+FTB2	10.51±0.63	11.88±1.38	13.36±1.00	15.55±1.22	5.09±0.08	5.19±0.04	6.21±0.87	7.44±0.86

*Individual readings n =3 for each plant were averaged and are presented with ±SD.

*Significant correlation was found between the root and shoot length at 0.05 level of significance;

FTB1 and FTB2 (Fluoride tolerant bacteria = Clostridium sp).

Table 4.4.2: Root and shoot fresh weight of *P. juliflora* plants grown in soil for 30, 60, 90 and 120 days in different F concentration viz. control, 25, 50, 75 and 100 mg kg^{-1} NaF soil after treatment with microbes FTB1 and FTB2

Biomass content								
Root fresh weight (gm)					**Shoot fresh weight (gm)**			
Treatment	**30 days**	**60 days**	**90 days**	**120 days**	**30 days**	**60 days**	**90 days**	**120 days**
Control	0.241±0.00	0.172±0.00	0.202±0.00	0.232±0.00	0.233±0.00	0.242±0.00	0.243±0.01	0.262±0.01
Control 25	0.152±0.00	0.163±0.00	0.193±0.00	0.194±0.02	0.215±0.01	0.233±0.00	0.245±0.00	0.263±0.01
25 + FTB1	0.163±0.00	0.173±0.01	0.216±0.02	0.257±0.03	0.274±0.00	0.284±0.00	0.315±0.01	0.335±0.01
Control 50	0.165±0.00	0.175±0.00	0.184±0.00	0.194±0.02	0.223±0.00	0.235±0.00	0.244±0.00	0.255±0.01
50+ FTB1	0.172±0.00	0.184±0.00	0.215±0.02	0.254±0.00	0.272±0.00	0.283±0.00	0.323±0.00	0.325±0.00
Control 75	0.171±0.00	0.187±0.00	0.195±0.01	0.203±0.02	0.233±0.00	0.234±0.00	0.253±0.00	0.266±0.00
75 + FTB1	0.182±0.00	0.193±0.00	0.217±0.03	0.264±0.00	0.286±0.00	0.294±0.00	0.333±0.00	0.346±0.00
Control 100	0.184±0.00	0.195±0.00	0.205±0.01	0.214±0.01	0.246±0.00	0.245±0.00	0.266±0.00	0.276±0.00
100 + FTB1	0.232±0.00	0.203±0.00	0.233±0.03	0.284±0.00	0.293±0.00	0.325±0.00	0.336±0.00	0.356±0.00
Control 25	0.142±0.00	0.155±0.00	0.164±0.00	0.174±0.02	0.206±0.00	0.234±0.01	0.237±0.00	0.236±0.00
25 + FTB2	0.154±0.00	0.174±0.01	0.196±0.00	0.205±0.00	0.215±0.00	0.214±0.00	0.266±0.00	0.263±0.01

Control 50	0.164±0.00	0.173±0.01	0.175±0.01	0.175±0.01	0.215±0.00	0.234±0.00	0.245±0.00	0.247±0.00
50 + FTB2	0.165±0.00	0.186±0.01	0.206±0.00	0.217±0.00	0.217±0.00	0.225±0.00	0.277±0.00	0.277±0.00
Control 75	0.172±0.00	0.176±0.00	0.193±0.00	0.197±0.01	0.217±0.00	0.245±0.00	0.255±0.00	0.255±0.00
75 + FTB2	0.181±0.00	0.186±0.01	0.215±0.00	0.224±0.01	0.226±0.00	0.237±0.00	0.274±0.00	0.285±0.00
Control 100	0.184±0.00	0.196±0.00	0.206±0.00	0.215±0.03	0.227±0.00	0.244±0.00	0.265±0.00	0.254±0.00
100 + FTB2	0.191±0.00	0.193±0.00	0.215±0.00	0.235±0.00	0.229±0.00	0.234±0.00	0.284±0.00	0.294±0.00
Control 25	0.117±0.01	0.138±0.00	0.145±0.00	0.143±0.00	0.164±0.00	0.164±0.00	0.176±0.00	0.184±0.00
25+ FTB1+FTB2	0.142±0.00	0.144±0.00	0.155±0.00	0.167±0.00	0.166±0.00	0.175±0.00	0.184±0.00	0.195±0.00
Control 50	0.144±0.01	0.146±0.00	0.145±0.00	0.153±0.00	0.177±0.00	0.183±0.00	0.194±0.00	0.203±0.01
50+ FTB1+FTB2	0.151±0.01	0.154±0.00	0.166±0.00	0.167±0.00	0.198±0.00	0.194±0.00	0.205±0.00	0.222±0.00
Control 75	0.155±0.02	0.146±0.00	0.156±0.00	0.152±0.00	0.185±0.00	0.187±0.00	0.194±0.00	0.205±0.01
75 + FTB1+FTB2	0.165±0.01	0.156±0.00	0.156±0.00	0.176±0.00	0.197±0.00	0.203±0.01	0.215±0.00	0.235±0.01
Control 100	0.169±0.01	0.157±0.00	0.167±0.00	0.164±0.00	0.185±0.00	0.195±0.00	0.195±0.00	0.216±0.01
100 + FTB1+FTB2	0.169±0.01	0.165±0.00	0.164±0.00	0.172±0.00	0.192±0.00	0.217±0.01	0.237±01	0.240±0.00

*Individual readings n =3 for each plant were averaged and are presented with ±SD.

*Significant correlation was found between the root and shoot fresh weight at 0.05 level of significance; FTB1 and FTB2 (Fluoride tolerant bacteria = Clostridium sp).

Table 4.4.3: Root and shoot dry weight of *P. juliflora* plants grown in soil for 30, 60, 90 and 120 days in different F concentration viz. control, 25, 50, 75 and 100 mg kg^{-1} NaF soil after treatment with microbes FTB1 and FTB2

Biomass content								
Root dry weight (gm)					**Shoot dry weight (gm)**			
Treatment	**30 days**	**60 days**	**90 days**	**120 days**	**30 days**	**60 days**	**90 days**	**120 days**
Control	0.165±0.00	0.152±0.00	0.163±0.00	0.183±0.00	0.215±0.01	0.234±0.00	0.257±0.00	0.254±0.00
Control 25	0.147±0.00	0.142±0.00	0.154±0.00	0.173±0.00	0.186±0.00	0.205±0.01	0.214±0.00	0.225±0.00
25+ FTB1	0.155±0.00	0.173±0.00	0.182±0.00	0.195±0.00	0.246±0.00	0.243±0.00	0.266±0.00	0.275±0.01
Control 50	0.154±0.00	0.145±0.00	0.155±0.00	0.176±0.00	0.206±0.00	0.213±0.00	0.336±0.01	0.245±0.00
50+ FTB1	0.155±0.00	0.173±0.00	0.185±0.00	0.194±0.00	0.256±0.00	0.264±0.00	0.277±0.00	0.285±0.01
Control 75	0.155±0.00	0.155±0.00	0.167±0.00	0.187±0.00	0.215±0.01	0.235±0.00	0.247±0.01	0.256±0.00
75+ FTB1	0.157±0.00	0.184±0.00	0.195±0.00	0.195±0.00	0.265±0.00	0.267±0.00	0.285±0.00	0.297±0.01
Control 100	0.165±0.00	0.154±0.00	0.177±0.00	0.195±0.00	0.235±0.03	0.273±0.00	0.285±0.00	0.265±0.00
100 + FTB1	0.164±0.00	0.184±0.00	0.218±0.00	0.225±0.00	0.286±0.00	0.284±0.00	0.294±0.00	0.316±0.02
Control 25	0.144±0.00	0.157±0.00	0.157±0.00	0.174±0.00	0.197±0.00	0.204±0.02	0.236±0.00	0.247±0.00
25+ FTB2	0.156±0.00	0.164±0.00	0.175±0.00	0.186±0.01	0.234±0.00	0.245±0.00	0.267±0.00	0.267±0.00
Control 50	0.145±0.00	0.155±0.00	0.165±0.00	0.166±0.00	0.216±0.00	0.215±0.00	0.246±0.01	0.258±0.00
50+ FTB2	0.155±0.00	0.167±0.00	0.175±0.00	0.187±0.00	0.236±0.00	0.246±0.00	0.277±0.01	0.287±0.00
Control 75	0.155±0.00	0.158±0.00	0.164±0.00	0.175±0.00	0.215±0.00	0.216±0.01	0.258±0.01	0.267±0.00
75 + FTB2	0.166±0.00	0.175±0.00	0.184±0.00	0.194±0.00	0.246±0.00	0.257±0.00	0.284±0.01	0.296±0.00

Control 100	0.156±0.00	0.165±0.00	0.176±0.00	0.184±0.00	0.226±0.00	0.235±0.00	0.267±0.01	0.278±0.00
100 + FTB2	0.167±0.00	0.183±0.00	0.196±0.00	0.215±0.00	0.257±0.00	0.268±0.00	0.294±0.01	0.307±0.00
Control 25	0.137±0.00	0.145±0.00	0.146±0.00	0.156±0.00	0.176±0.00	0.174±0.00	0.187±0.00	0.197±0.00
25+ FTB1+FTB2	0.146±0.00	0.147±0.00	0.156±0.00	0.164±0.00	0.228±0.01	0.246±0.00	0.257±0.00	0.268±0.00
Control 50	0.147±0.00	0.159±0.00	0.165±0.01	0.175±0.00	0.185±0.00	0.184±0.00	0.203±0.01	0.218±0.00
50+ FTB1+FTB2	0.157±0.00	0.156±0.00	0.164±0.01	0.186±0.00	0.225±0.00	0.249±0.00	0.257±0.00	0.267±0.00
Control 75	0.158±0.00	0.165±0.00	0.165±0.00	0.175±0.00	0.195±0.00	0.204±0.01	0.218±0.00	0.227±0.00
75+ FTB1+FTB2	0.156±0.00	0.169±0.00	0.177±0.00	0.184±0.00	0.236±0.00	0.256±0.00	0.264±0.00	0.278±0.00
Control 100	0.158±0.00	0.160±0.00	0.178±0.00	0.186±0.00	0.195±0.00	0.217±0.01	0.226±0.00	0.236±0.00
100+ FTB1+FTB2	0.160±0.00	0.173±0.00	0.393±0.00	0.185±0.00	0.242±0.00	0.258±0.01	0.268±0.00	0.287±0.00

*Individual readings n =3 for each plant were averaged and are presented with ±SD.

*Significant correlation was found between the root and shoot dry weight at 0.05 level of significance; FTB1 and FTB2 (Fluoride tolerant bacteria = Clostridium sp).

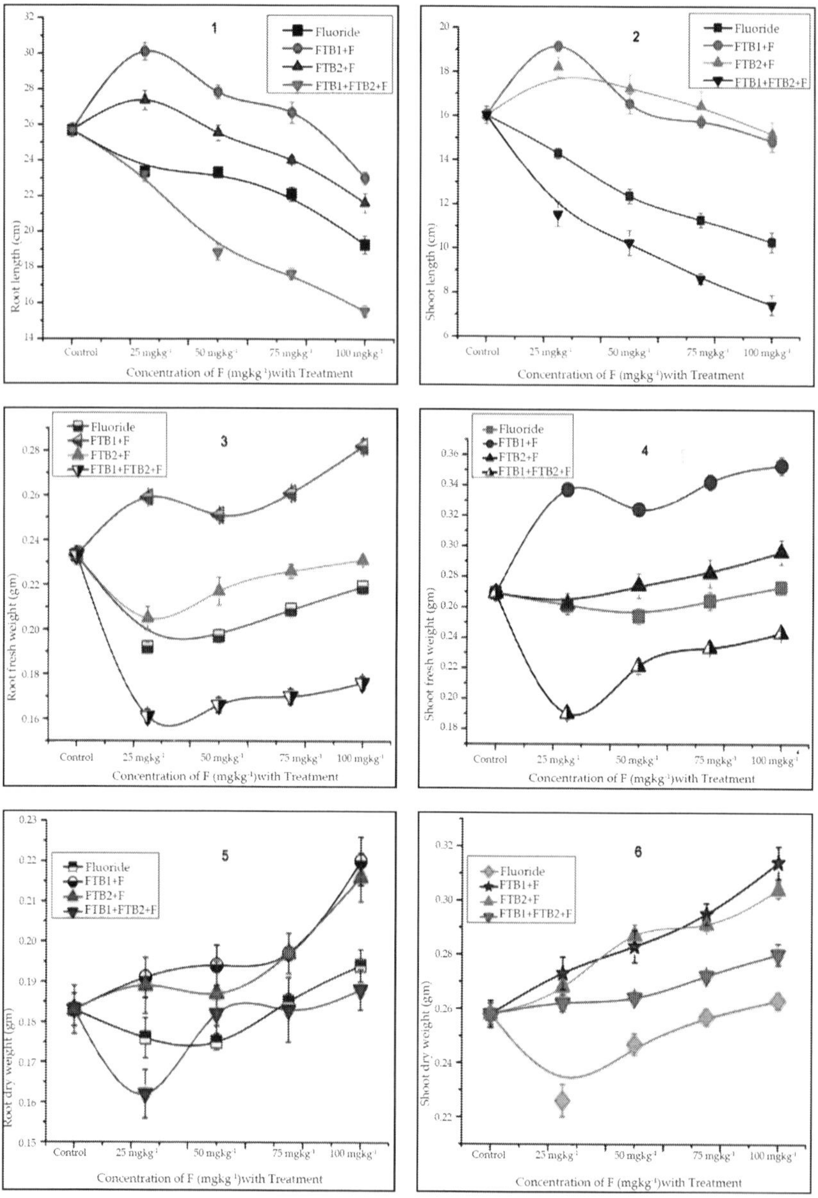

Figure 4.7: Effect of F-tolerant bacteria (FTB1, FTB2and FTB1+FTB2) on biomass root length (1), shoot length (2), root fresh weight (3), shoot fresh weight (4), root dry weight (5) and shoot dry weight (6) of *P. juliflora* plants grown in soil under different concentration of F, viz., control, 25, 50, 75 and 100 mg kg^{-1}NaF for 120 days. SE denoted by vertical bars.

4.6 ORGAN-WISE F UPTAKE

F accumulated by *P. juliflora* seedlings 30.40 mg kg-1 NaF in root, 41.21 mg kg^{-1}NaF in shoot and 28.39 mg kg-1 remaining in soil at 100 mg kg-1 NaF+FTB1(Figure 4.8 Table 4.5 and 4.6). The inoculation of *Clostridium sp.* significantly increased ($p \leq 0.05$) F concentration in shoots. The microbial-induced promoting effect on F accumulation in roots was observed (Figure 4.8).

The total F contents in shoots and roots of plants are shown in (Table 4.5). It was almost two times higher accumulation than exposed to F only, however, the F accumulation in roots was low in comparison to the shoot (Figure 4.8).

We found an overall enhancement in F accumulations as compared to the control group, while the F accumulation in shoots was higher affected and more hyperaccumulation as compared to the root. However, *Clostridium sp.*mainly enhanced the accumulation of F in both shoots and roots at 100 mg kg^{-1}NaF+FTB1.

The bioaccumulation factor (BF) values increased as (1.45).*P. juliflora* showed a translocation factor (T.F) was (1.35) at 100 mg kg-1 NaF+FTB1 (Figure s 4.9 and Table 4.7). Metal-resistant bacteria can promote plant growth and enhance metal uptake by hyperaccumulator plants (Zaidi *et al.*, 2006; Abou-Shanab *et al.*, 2006; Ashok *et al.*, 2010). Pot experiments confirmed that FTB1 facilitated plant growth and also increased plant biomass.

To the best of our information, this is the first research report on the effect of *Clostridium sp.* on F solubilization in soil and accumulation by *P.juliflora* in pot experiments. Kimura *et al.* (2004) in their study reported a higher amount of F accumulation in roots than shoots of onion (*Allium cepa L.*).

The ratio greater than 1 means hyperaccumulation of F in a plant while a ratio of less than 1 is a hyperaccumulator (Gupta and Banerjee, 2011). Combination of FTB1 and FTB2 strains was not effective as compared to individual performance.

Several species of *Bacillus*and *Microbacterium*were heavy metals tolerant(Anandham *et al.*, 2008). Previous research reported that Arsenic (As) uptake by *P. deltoids, P. vita* was enhanced under As-polluted soils due to the inoculation of As-resistant bacteria (Broadhurst *et al.*, 2004; Ghosh *et al.*, 2011).

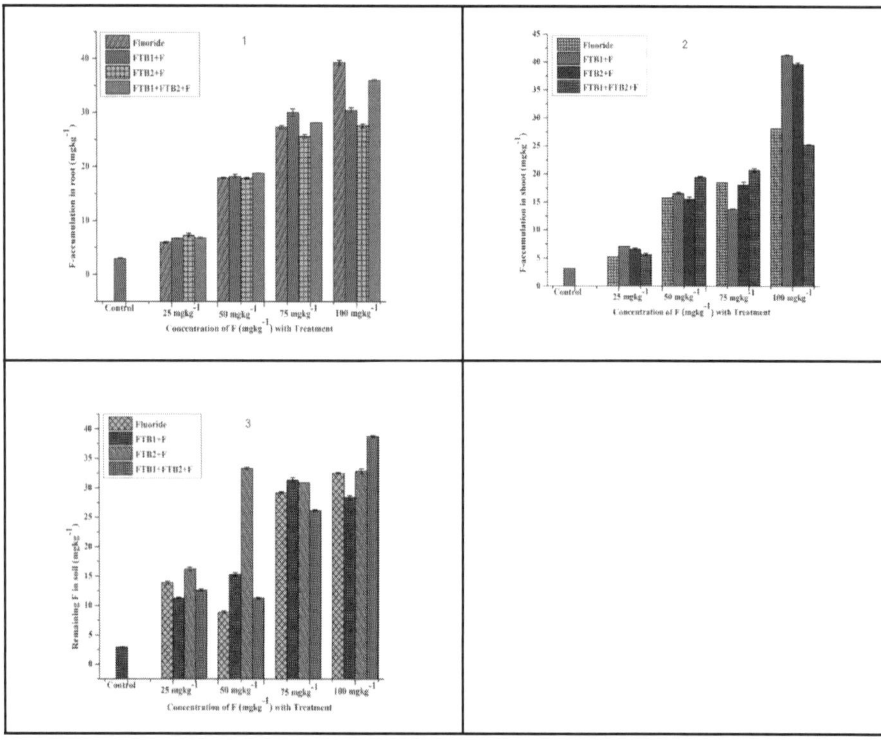

Figure 4.8: Effect of F-tolerant bacteria (FTB1, FTB2, and FTB1+FTB2) on root (1), shoot (2) uptake and remaining F in soil (3) values of *P. juliflora* plants grown in soil under different concentration of F, viz., control, 25, 50, 75 and 100 mg kg^{-1}NaF for 120 days. SE denotes vertical bars.

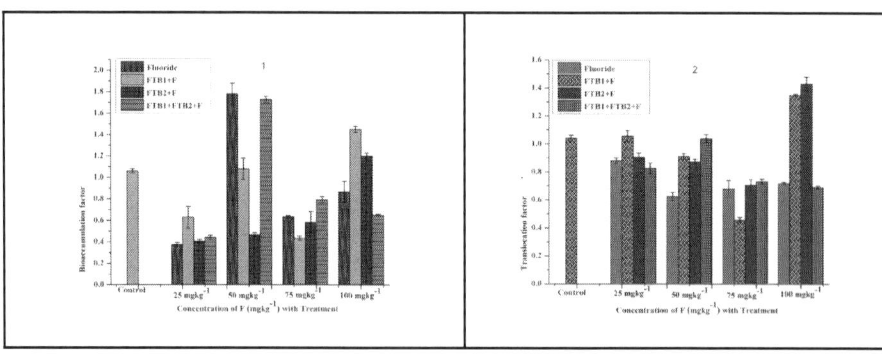

Figure 4.9: Effect of F-tolerant bacteria (FTB1, FTB2 and FTB1+FTB2) on (1) bioaccumulation (2) translocation factor values of *P. juliflora* plants grown in soil under different concentration of F, viz., control, 25, 50, 75 and 100 mg kg-1 NaF for 120 days. SE denotes vertical bars.

Table 4.5: Fluoride accumulation in root and shoot of *P. juliflora* seedlings were grown in soil for 30, 60, 90 and 120 days in different concentrations of F viz. control, 25, 50, 75 and 100 mg kg^{-1} NaF after treatment with microbes FTB1 and FTB2

Fluoride content (mg kg^{-1})								
Root uptake					**Shoot uptake**			
Treatment	**30 days**	**60 days**	**90 days**	**120 days**	**30 days**	**60 days**	**90 days**	**120 days**
Control	1.23±0.01	1.53±0.75	2.26±0.05	2.96±0.44	1.30±0.06	2.11±0.19	2.77±0.57	3.11±0.19
Control 25	4.66±0.11	4.81±0.35	5.73±0.89	5.92±0.13	3.29±0.06	3.40±0.52	4.18±0.31	5.21±0.57
25+ FTB1	5.71±0.88	5.51±0.45	6.10±0.38	6.69±0.12	4.20±0.17	4.22±0.19	4.77±0.76	7.06±2.43
Control 50	13.73±0.71	14.40±1.11	15.81±0.64	25.33±1.15	10.92±0.51	11.36±1.05	13.07±1.09	15.81±1.28
50+FTB1	15.14±0.25	15.66±0.79	17.66±1.11	18.17±0.71	12.14±0.12	13.03±0.89	14.18±1.09	16.55±1.92
Control 75	21.51±0.44	23.18±1.95	25.51±0.89	27.29±0.64	17.92±0.88	18.10±0.58	19.18±0.31	18.51±0.44
75+ FTB1	22.85±1.28	22.92±0.81	25.32±1.63	29.99±3.97	18.51±0.44	19.29±0.50	20.44±0.98	13.70±9.75
Control 100	31.44±1.06	32.88±2.91	36.92±1.59	39.29±0.45	23.68±0.61	25.09±0.60	25.96±0.90	28.18±1.02
100+ FTB1	34.62±0.73	35.40±1.29	37.36±1.40	30.40±4.55	26.10±0.95	27.03±0.50	26.95±0.70	41.21±1.53
Control 25	4.07±0.12	4.73±0.63	4.81±0.45	6.14±0.70	4.11±0.19	4.22±0.19	4.55±0.38	5.50±0.43
25+ FTB2	5.30±0.06	5.55±0.28	6.17±0.51	7.25±0.42	3.82±0.71	3.36±0.31	4.18±1.09	6.55±1.92
Control 50	12.40±0.52	13.99±1.63	15.81±0.64	17.24±0.64	10.81±0.51	11.92±1.54	13.81±1.02	16.77±1.53
50 + FTB2	14.40±0.12	14.84±0.63	16.66±1.05	17.80±1.00	11.96±0.45	12.77±1.15	13.73±0.51	15.51±2.04

Control 75	20.77±1.07	22.70±2.10	25.07±1.28	27.29±0.64	18.80±0.06	18.55±0.48	18.85±0.12	21.14±1.85
75+ FTB2	21.10±0.00	23.29±2.16	25.07±0.64	25.56±1.22	16.32±0.76	16.18±0.64	16.92±0.64	18.06±0.70
Control 100	31.74±0.45	33.70±2.23	35.88±0.57	38.03±1.28	25.62±1.98	24.47±1.16	26.10±0.77	27.96±0.83
100 + FTB2	33.81±0.32	33.33±1.05	35.07±1.28	27.54±1.15	22.96±0.64	24.36±1.85	25.92±0.83	39.63±4.30
Control 25	3.11±0.11	4.18±0.94	5.92±1.40	7.18±0.31	2.80±0.25	3.58±0.74	4.47±0.06	5.43±0.76
25 + FTB1+FTB2	4.07±0.13	4.44±0.50	5.33±0.57	6.77±0.69	3.25±0.25	3.47±0.06	3.96±0.90	5.59±1.02
Control 50	11.37±0.54	12.92±1.64	14.81±0.83	16.36±0.51	11.18±0.71	13.92±1.75	16.18±1.28	17.66±0.00
50+ FTB1+FTB2	16.18±0.22	16.10±0.40	16.92±0.64	18.75±1.11	14.29±0.31	15.32±0.57	16.44±1.35	19.48±2.56
Control 75	22.77±0.48	23.96±1.58	25.81±0.83	27.36±0.51	20.62±3.21	25.58±1.22	23.29±0.79	25.11±0.96
75+ FTB1+FTB2	19.17±0.61	22.18±2.22	24.77±0.76	28.14±0.57	15.88±0.57	18.03±1.28	19.14±0.64	20.73±1.47
Control 100	31.81±0.73	33.81±1.18	34.84±0.70	35.51±0.12	22.81±0.73	23.33±0.29	24.88±1.01	27.84±1.59
100 + FTB1+FTB2	31.55±1.15	32.33±2.16	34.92±0.64	36.03±0.46	21.59±0.64	23.07±0.64	24.14±0.94	25.21±1.15

*Individual readings n =3 for each plant were averaged and are presented with ±SD.

*Significant correlation was found between Fluoride content (root and shoot) at 0.05 level of significance;

FTB1 and FTB2 (Fluoride tolerant bacteria = Clostridium sp).

Table 4.6: Remaining F concentration in the soil after harvesting the plantlets of *P. juliflora* for 30, 60, 90 and 120 days in different concentrations of F viz. control, 25, 50, 75, and 100 mg kg^{-1} NaF after treatment with microbes FTB1 and FTB2

Fluoride content (mg kg^{-1})				
Treatment	**30 days**	**60 days**	**90 days**	**120 days**
Control	1.60±0.00	1.70±0.20	2.70±0.06	2.93±0.65
Control 25	17.05±0.10	16.79±1.00	15.09±0.20	13.87±0.10
25+ FTB1	15.09±0.10	15.47±0.00	14.13±0.20	11.25±0.10
Control 50	25.35±0.00	24.24±1.00	21.12±0.10	8.86±0.10
50+ FTB1	22.72±2.00	21.31±0.00	18.16±0.60	15.28±0.10
Control 75	35.57±0.00	33.72±1.00	30.31±0.80	29.20±0.20
75+ FTB1	33.63±0.00	32.79±0.00	29.24±0.00	31.37±0.10
Control 100	44.88±0.40	42.03±1.00	37.12±0.30	32.53±0.10
100 + FTB1	60.72±0.00	37.57±1.00	35.69±0.30	28.39±0.00
Control 25	16.82±0.20	16.05±0.00	15.64±0.20	13.36±0.20
25 + FTB2	15.88±0.30	16.09±0.00	14.65±0.10	16.20±0.00
Control 50	26.79±0.00	24.09±1.00	20.32±0.00	15.94±0.20
50 + FTB2	23.64±0.40	22.39±0.00	19.61±0.10	33.31±0.20
Control 75	35.43±0.10	33.75±2.00	31.08±0.00	26.17±0.00
75 + FTB2	37.58±0.20	35.53±1.00	33.01±0.20	30.93±0.20
Control 100	42.64±0.10	41.83±1.00	38.02±0.10	34.01±0.20
100 + FTB2	56.77±0.10	42.31±0.00	39.01±0.10	32.93±0.10
Control 25	19.09±0.40	17.24±1.00	14.61±0.00	12.39±0.20
25 + FTB1+FTB2	17.67±0.20	17.09±1.00	15.71±0.10	12.64±0.10
Control 50	22.55±0.20	23.16±0.00	19.00±0.10	15.98±0.10
50+ FTB1+FTB2	19.53±0.10	18.58±1.00	16.64±0.20	11.27±0.20
Control 75	31.61±0.00	25.46±1.00	25.09±0.10	22.53±0.10
75 + FTB1+FTB2	44.95±0.60	34.79±0.20	31.09±0.10	26.13±0.20

Control 100	55.38±1.00	42.86±0.20	40.33±0.10	36.65±0.20
100+ FTB1+FTB2	46.66±1.00	55.40±0.30	40.64±0.10	38.76±0.10

*Individual readings n =3 for each plant were averaged and are presented with ±SD.

*Significant correlation was found between Fluoride content (root and shoot) at 0.05 level of significance;

FTB1 and FTB2 (Fluoride tolerant bacteria = Clostridium sp).

Table 4.7: Translocation and bioaccumulation factor values (shoot/soil concentration ratio) of *P. juliflora* plants grown in soil under different concentration of F viz. control, 25, 50, 75 and 100 mg kg^{-1} NaF soil for 30, 60, 90 and 120 days after treatment with microbes FTB1 and FTB2.

Fluoride content								
Translocation					**Bioaccumulation**			
Treatment	**30 days**	**60 days**	**90 days**	**120 days**	**30 days**	**60 days**	**90 days**	**120 days**
Control	1.05	1.37	1.22	1.04	0.81	1.24	1.02	1.06
Control 25	0.70	0.70	0.72	0.88	1.93	0.20	0.27	0.37
25 + FTB1	0.73	0.76	0.78	1.05	0.27	0.27	0.33	0.62
Control 50	0.79	0.78	0.82	0.62	0.43	0.46	0.61	1.78
50+ FTB1	0.80	0.83	0.80	0.91	0.53	0.61	0.78	1.08
Control 75	0.83	0.78	0.75	0.67	0.50	0.53	0.63	0.63
75+ FTB1	0.81	0.84	0.80	0.45	0.55	0.58	0.69	0.43
Control 100	0.75	0.76	0.70	0.71	0.52	0.59	0.69	0.86
100+ FTB1	0.75	0.76	0.72	1.35	0.42	0.71	0.75	1.45
Control 25	1.00	0.89	0.94	0.89	0.24	0.26	0.29	0.41
25 + FTB2	0.72	0.60	0.67	0.90	0.24	0.20	0.28	0.40
Control 50	0.87	0.85	0.87	0.97	0.40	0.49	0.67	1.05
50 + FTB2	0.83	0.86	0.82	0.87	0.50	0.57	0.70	0.46
Control 75	0.90	0.81	0.75	0.77	0.53	0.54	0.60	0.80
75+ FTB2	0.77	0.69	0.67	0.70	0.43	0.45	0.51	0.58
Control 100	0.80	0.72	0.72	0.73	0.60	0.58	0.68	0.82
100+ FTB2	0.67	0.73	0.73	1.43	0.40	0.57	0.66	1.20
Control 25	0.90	0.85	0.75	0.75	0.14	0.20	0.30	0.43
25+ FTB1+FTB2	0.79	0.78	0.74	0.82	0.18	0.20	0.25	0.44
Control 50	0.98	1.07	1.09	1.07	0.49	0.60	0.85	1.10

50+ FTB1+FTB2	0.88	0.95	0.97	1.03	0.73	0.82	0.98	1.72
Control 75	0.90	1.06	0.90	0.91	0.65	1.00	0.89	1.11
75+ FTB1+FTB2	0.82	0.81	0.77	0.73	0.35	0.51	0.61	0.79
Control 100	0.71	0.69	0.71	0.78	0.41	0.54	0.61	0.75
100+ FTB1+FTB2	0.68	0.71	0.69	0.69	0.46	0.41	0.60	0.65

*Significant correlation was found at 0.05 level of significance;

FTB1 and FTB2 (Fluoride tolerant bacteria = Clostridium sp).

4.7 EFFECT OF MICROBIAL CONSORTIUM (P.F AND P.A) ON GROWTH PARAMETERS

The germination percentage was 50 percent in case of P.F treatment and 40 percent at 100 mg kg^{-1} NaF with P.A. The effect of microbes P.F and P.A on plant growth was observed (Figure 4.10 and Table 4.8.1-4.8.4) under different concentrations of F.

(1)

(2)

Plate 4.9.B: Effect of P.F and P.A on plant growth for 120 days of pot experiment with different concentrations of NaF viz. control, 25, 50, 75 and 100 mg kg^{-1} soil.

(1) P.F (*Pseudomonas fluorescens*)

(2) P.A (*Pseudomonas aeruginosa*)

The plant showed promoting effects on the root, shoot length (Figure 4.10- 1 and 2) and biomass production (Figure 4.10-3, 4, 5, 6) of *P. juliflora* under microbes treatment (Figure 4.10). Although the length and biomass of plant tissues were significantly ($p \leq 0.05$) affected at 100 mg kg^{-1} NaF. *P. juliflora* did not show any toxic symptom when plant under normal stress condition but if we were to go above more than 100 mg kg^{-1}then it showed chlorosis and necrosis effect on the plant due to more accumulation in tissue part and no translocation of F.

Then, the P.F and P.A significantly ($p \leq 0.05$) increased both, root length (19.36cm)and the root biomass (fresh weight from 0.288g and dry weight from 0.293g).

The P.F and P.A both performed better in case of root parameters and P.F promoted shoot length (18.53cm) and the shoot biomass (fresh weight from 0.349 g and dry weight from 0.322g) of plants under F stress. However, P.A increased both the root length (23.96 cm) and the root biomass (fresh weight from 0.284 g and dry weight from 0.283 g) in comparison to the control group.

Meanwhile, P.A increased both the shoot length (16.03 cm) and the shoot biomass (fresh weight from 0.346 g and dry weight from 0.291 g) was significantly increased.

However, F stress had decreased root length, shoot length and also affected plant biomass at 100 mg kg-1 F stresses.

However, microbial consortium treatment has increased plant biomass as compared to control due to the bioavailability of mineral content in soil (Ghosh *et al*., 2011). The P.F have a tendency to increase F mobility in soil therefore, it formed a complex with (Ca, Mg, P, Al, Fe and Cu) because F has more electronegativity power to interact with other metals. However, these phenomena have important phases for enhanced plant biomass.

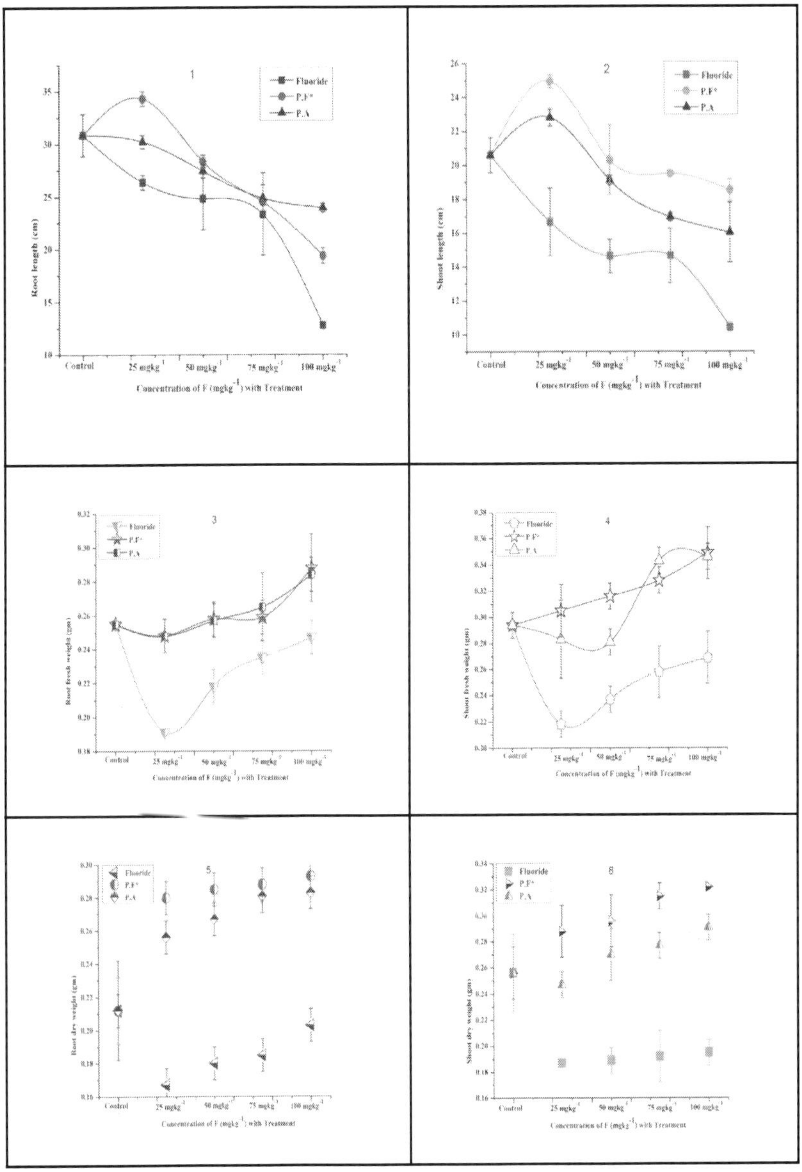

Figure 4.10: Effect of the microbial consortium on plant growth and biomass under F-contaminated soil for 120 days (1-root length, 2-shoot length,3-root dry weight, 4-shoot dry weight, 5-root fresh weight, 6-shoot fresh weight and P.F- *Pseudomonas fluorescens*, P.A- *Pseudomonas aeruginosa* significant value*).

4.8 EFFECT OF MICROBIAL CONSORTIUM (P.F AND P.A) ON CHLOROPHYLL CONTENTS

The chlorophyll content of *P. juliflora* leaves was reduced under F stress as compared to control but increased when microbes treatment has given as Chl a (3.329mgg^{-1}fw), Chl b (1.101mgg^{-1}fw), total Chl a+b (4.431mgg^{-1}fw) and Chl a/b (3.021mgg^{-1}fw) are shown in (Figure 4.11 and Table 4.9.1-4.9.2). Heavy metal stresses are known to be interfering in chlorophyll contents through the direct inhibition of enzyme activity which affects the plant nutrients. The chlorophyll substance was reduced under copper contaminated soil in *Triticum aestivum* and *Brassica oleracea*(Chatterjee, 2000). In previous research, it was found out that chlorophyll biosynthesis reduced and inhibited chlorophyll (Saini *et al.*, 2012). Fe effect on chlorophyll content was highly reduced at the transient phase (stationary state) approximately 90 days. After that crucial period, the plant was not affected under given P.F treatment as compared to control under F stress after 120 days.

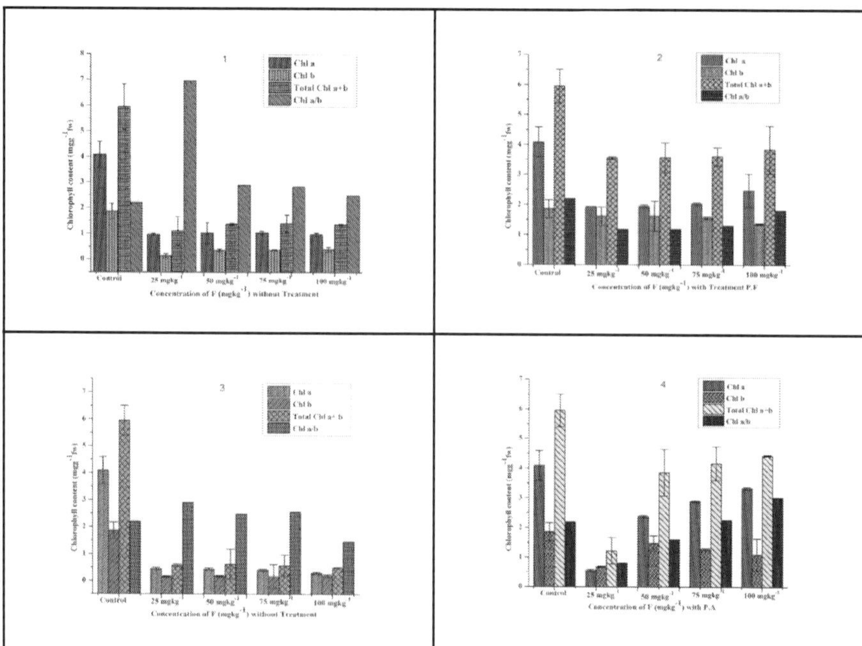

Figure 4.11: Effect of the microbial consortium on chlorophyll contents of *P. juliflora* underF- stress for 120 days (1,2-P.F- *Pseudomonas fluorescens,* 3,4-P.A- *Pseudomonas aeruginosa* where significant value*).

4.9 EFFECT OF MICROBIAL CONSORTIUM (P.F AND P.A) ON ANTIOXIDANT ENZYME ACTIVITY

Fluoride as a strong oxidant can directly induce the generation of reactive oxygen species (ROS) and oxidative stress in plants. Fluoride could inhibit or activate enzymes which protect the organism against imbalance between ROS and antioxidants.

The antioxidant enzyme activities(POD, CAT and SOD) in the leaves of control and treated with differentconcentration of Fexposed to germplasm of *P.juliflora* with the microbial consortium(P.F and P.A) were found. The antioxidant activity increased with an increase in F concentration after 30 days and continuously increased up to 90 days, then decreased gradually after 120days of treatment. In leaves, the antioxidant enzyme activities were significantly higher at 100mg kg-1 NaF (0.214), (0.180) and (0.170 $\mu g g^{-1}$ fw) treated plant of *P.juliflora* for 120 days (Figure 4.12 and Table 4.10.1 and 4.10.2).

The hydrogen peroxide is toxic for plants but it behaves both as oxidant and reductant. The hyperactivity of POD under metal stress indicated the scavenging activity of H_2O_2 generated through the activity of photorespiration in plant cells. Therefore, an increase in POD activity protects plants from toxic effects (Ali *et al.*, 2003).

Catalase activity scavenges H_2O_2 by breaking it directly to form water and oxygen. Similar study increases in CAT activity were also obtained with an increase in salinity amount in *Triticum aestivum* (Low and Merida, 1996; Heidari, 2009).

The activity of enzymes increased with increase F concentration as well as doses of rhizobacteria.

Among these enzymes, superoxide dismutase (SOD) is the first line of defense against ROS, dismutating O^{-2} to an oxygen molecule and H_2O_2 (Gill *et al.*, 2011).

The oxidative stress induced by F in plants appears to be an indirect effect of fluoride toxicity to plant (Kumar *et al.*, 2009). The present study suggested that microbe P.F and P.A could help host plants to cope with F toxicity.

However, suggesting that different antioxidant enzymes showed different sensitivity to F contamination and it was highly more sensitive to F ions.

This phenomenon could be reasonably explained by Azcon *et al.* (2009)

who suggested that microbe symbiosis around the root cells might provide protection against oxidative stress and removal of ROS from plant cells.

Numerous studies have indicated that microbes increase the antioxidant activity, POD, SOD and CAT enzymes are indispensable compounds for the cellular defense strategy towards the higher ROS production in plant cells (Giannakoule *et al.*, 2010; Yang *et al.*, 2015).

The mechanism of metabolic pathways is still poorly understood which requires further research in case of plant growth- promoting bacteria.

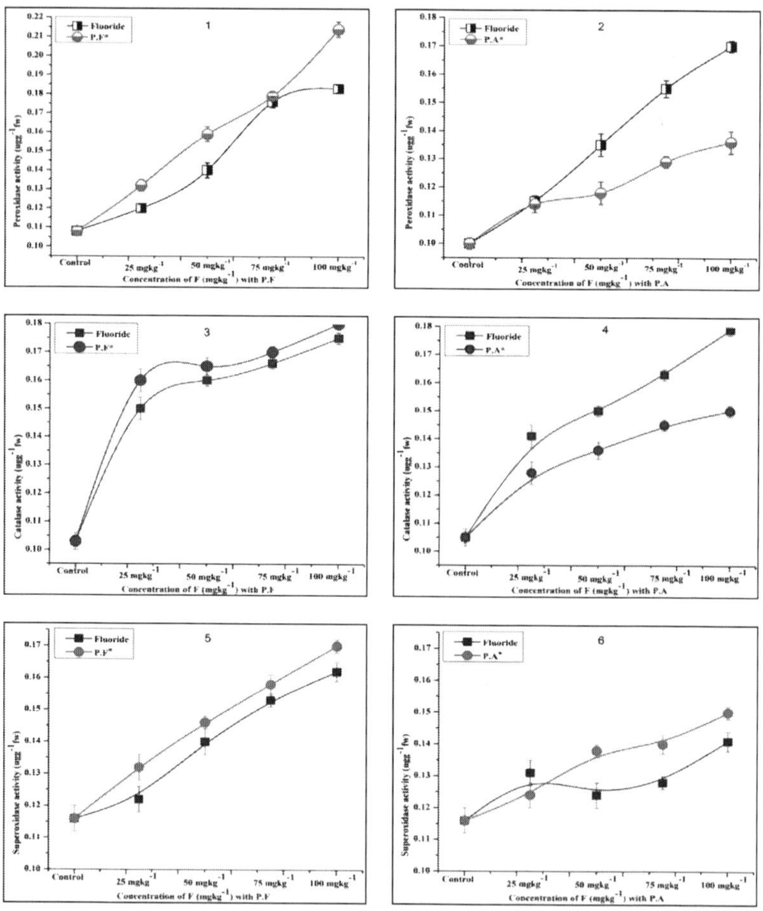

Figure 4.12: Effect of the microbial consortium on anti-oxidantactivity of *P. juliflora* under F- stress for 120 days (1,2-peroxidase, 3,4-catalase and 5,6-superoxidaseP.F-*Pseudomonas fluorescens,* P.A- *Pseudomonas aeruginosa* where significant value*).

4.10 EFFECT OF MICROBIAL CONSORTIUM (P.F AND P.A) ON F ACCUMULATION

The *P. juliflora* seedlings accumulated F (40.55 mg kg^{-1}NaF soil) in root dry tissue, (44.18 mg kg^{-1}NaF soil) in shoot dry tissue and remaining F in soil (15.27 mg kg-1 at 100 mg kg^{-1}NaF soil) (Figure s 4.13 and 4.14).In the current study, we report that there was high accumulation in roots in comparison to shoots, than was translocated to shoots. The accumulation of F in root could be due to the fact that the soluble fluoride fraction in the soil is taken up passively by roots as the non-ionic form at low pH and translocation into the shoot (Horner and Bell, 1995). A significantly positive correlation was found between F concentration with microbial treatment and uptake of F at ($P \leq 0.05$) level of significance. The bioaccumulation factor (BF) values were 2.89 and 1.34 at different microbial consortium amendments (P.F and P.A) with various F concentrations. *P. juliflora* showed a translocation factor (T.F) 1.08 and 1.04 which are shown in (Figure s 4.14 and Table 4.14). The ratio of the translocation factor was greater than 1 mean hyperaccumulator plant (Gupta *et al.*, 2009). The result of the present investigation showed that microbial consortium had the high ability and increased the efficiency of F accumulation. Plant growth -promoting bacteria deserve a special attention because they can directly improve the phytoremediation process by increasing the F solubility through altering soil pH, release of chelators (e.g., organic acids, siderophores), and they also help to increase metal mobility for high accumulation in plant organ(Rajkumar *et al.*, 2010; Singh *et al.*, 2017).

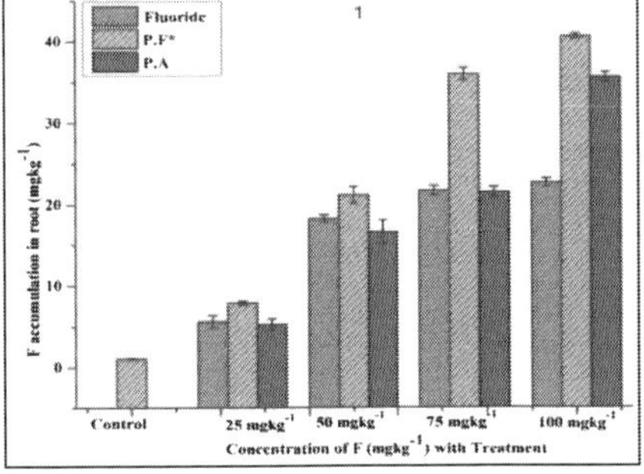

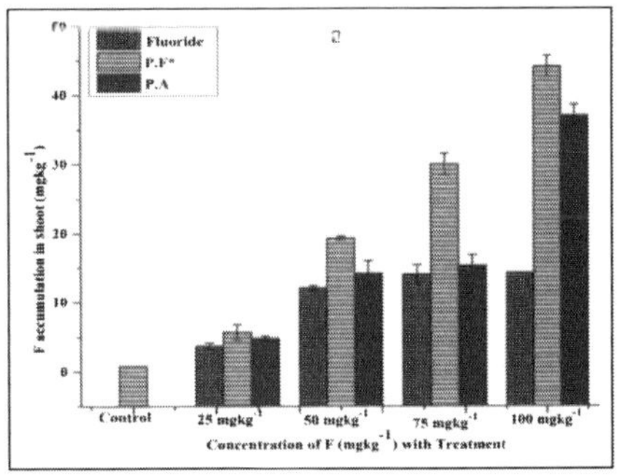

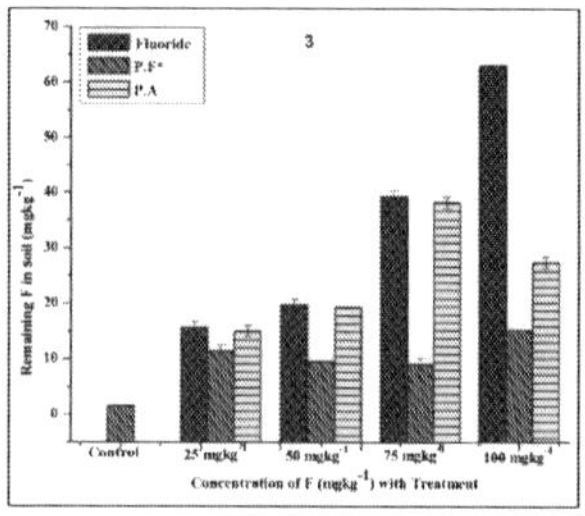

Figure 4.13: Effect of microbial consortium on root and shoot uptake under F stress for 120 days (1-root, 2- shoot uptake and 3-remaining F in soil, P.F- *Pseudomonas fluorescens* and P.A- *Pseudomonas aeruginosa* where significant value*).

Table 4.8.1: Biomass of P. juliflora plants grown in soil for 30 days in different F concentration viz. control, 25, 50, 75 and 100 mg kg^{-1} NaF soil after treatment with microbes P.F and P.A

Parameters (30 days)						
Treatment **RL(cm)**		**SL(cm)**	**RDW(gm)**	**SDW(gm)**	**RFW (gm)**	**SFW (gm)**
Control	12.40±0.34	8.75±0.00	0.14±0.01	0.220±0.01	0.152±0.00	0.256±0.00
Control 25	12.10±0.22	7.43±0.00	0.12±0.00	0.220±0.00	0.172±0.00	0.153±0.02
25 + P.F	26.80±1.05	9.33±0.25	0.15±0.00	0.240±0.01	0.165±0.00	0.246±0.00
Control 50	11.90±0.00	7.21±0.00	0.12±0.00	0.218±0.00	0.172±0.09	0.156±0.05
50+ P.F	15.26±0.30	9.13±0.15	0.14±0.00	0.237±0.02	0.177±0.00	0.257±0.00
Control 75	10.66±0.25	8.33±0.25	0.13±0.01	0.220±0.08	0.164±0.00	0.177±0.06
75+ P.F	12.20±0.26	7.23±0.10	0.13±0.01	0.236±0.01	0.188±0.00	0.264±0.00
Control 100	10.60±0.69	7.06±0.00	0.11±0.03	0.205±0.01	0.148±0.06	0.177±0.00
100 + P.F	12.66±0.98	6.26±0.30	0.11±0.01	0.215±0.00	0.184±0.00	0.284±0.01
Control 25	14.20±0.00	8.53±0.15	0.13±0.00	0.204±0.01	0.158±0.08	0.162±0.08
25+ P.A	22.10±0.10	8.60±0.50	0.15±0.00	0.216±0.00	0.168±0.00	0.238±0.00
Control 50	13.10±1.50	7.94±0.15	0.13±0.05	0.206±0.00	0.174±0.06	0.188±0.00
50 + P.A	18.23±0.20	6.20±0.25	0.14±0.00	0.218±0.01	0.166±0.00	0.239±0.07
Control 75	12.80±0.46	6.54±0.25	0.14±0.05	0.212±0.00	0.182±0.09	0.173±0.08

75+ P.A	16.20±0.20	6.33±0.11	0.14±0.01	0.263±0.02	0.179±0.00	0.243±0.07
Control 100	11.00±0.00	6.32±0.15	0.14±0.03	0.214±0.04	0.152±0.06	0.188±0.00
100+ P.A	15.60±0.32	7.33±0.30	0.14±0.02	0.265±0.00	0.181±0.00	0.251±0.06

*Individual readings n =3 for each plant were averaged and are presented with ±SD.

*Significant correlation was found between root length (RL), shoot length (SL), root,

shoot dry weight (RDW, SDW) and root, shoot fresh weight (RFW, SFW) at 0.05 level of significance;

P.F: Pseudomonas fluorescens; P.A: Pseudomonas aeruginosa.

Table 4.8.2: Biomass of *P. juliflora* plants grown in soil for 60 days in different F concentration viz. control, 25, 50, 75 and 100 mg kg^{-1} NaF soil after treatment with microbes P.F and P.A.

Parameters (60 days)						
Treat-ment	**RL(cm)**	**SL(cm)**	**RDW(gm)**	**SDW(gm)**	**RFW (gm)**	**SFW (gm)**
Control	22.86±0.50	11.33±0.22	0.23±0.00	0.24±0.00	0.273±0.01	0.282±0.00
Control 25	15.21±3.29	6.033±0.63	0.13±0.00	0.18±0.00	0.173±0.00	0.182±0.00
25+ P.F	29.33±2.30	21.86±0.57	0.19±0.01	0.27±0.00	0.185±0.00	0.276±0.00
Control 50	14.34±3.29	5.53±0.20	0.12±0.01	0.17±0.00	0.163±0.01	0.173±0.00
50 + P.F	16.33±1.27	16.13±1.43	0.19±0.01	0.28±0.00	0.192±0.01	0.287±0.00
Control 75	13.66±4.44	5.06±0.56	0.11±0.00	0.15±0.00	0.157±0.00	0.164±0.00
75 + P.F	15.43±0.40	15.60±1.90	0.20±0.02	0.28±0.00	0.213±0.00	0.298±0.00
Control 100	10.80±4.15	4.63±0.96	0.10±0.00	0.12±0.01	0.144±0.00	0.152±0.00
100+ P.F	12.63±4.56	12.53±3.44	0.20±0.01	0.29±0.00	0.220±0.00	0.307±0.01
Control 25	17.23±1.27	11.63±0.75	0.13±0.00	0.18±0.00	0.183±0.00	0.177±0.00
25+ P.A	22.96±1.32	13.26±0.04	0.19±0.01	0.22±0.00	0.183±0.01	0.267±0.00

Control 50	16.13±1.27	14.43±0.45	0.14±0.01	0.17±0.01	0.180±0.00	0.172±0.00
50 + P.A	19.46±2.54	17.96±1.27	0.20±0.01	0.23±0.00	0.183±0.00	0.284±0.00
Control 75	13.26±2.02	14.23±0.80	0.11±0.01	0.19±0.01	0.157±0.05	0.172±0.08
75+ P.A	16.90±1.55	21.16±2.13	0.19±0.02	0.24±0.00	0.198±0.01	0.285±0.00
Control 100	10.63±3.52	6.30±0.69	0.11±0.02	0.11±0.01	0.143±0.05	0.166±0.00
100+ P.A	14.83±2.65	25.03±0.63	0.21±0.02	0.28±0.05	0.232±0.00	0.302±0.01

*Individual readings n =3 for each plant were averaged and are presented with ±SD.

*Significant correlation was found between root length (RL), shoot length (SL), root, shoot dry weight (RDW, SDW) and root,

shoot fresh weight (RFW, SFW) at 0.05 level of significance;

P.F: Pseudomonas fluorescens; P.A: Pseudomonas aeruginosa.

Table 4.8.3: Biomass of *P. juliflora* plants grown in soil for 90 days in different F concentration viz. control, 25, 50, 75 and 100 mg kg^{-1} NaF soil after treatment with microbes P.F and P.A

Parameters (90 days)						
Treatment	**RL(cm)**	**SL(cm)**	**RDW(gm)**	**SDW(gm)**	**RFW (gm)**	**SFW (gm)**
Control	30.13±2.02	15.32±0.22	0.19±0.00	0.23±0.00	0.244±0.00	0.295±0.00
Control 25	24.93±0.57	6.53±0.23	0.14±0.00	0.19±0.00	0.182±0.00	0.193±0.00
25+ P.F	32.50±0.86	24.93±2.67	0.22±0.00	0.27±0.00	0.233±0.00	0.297±0.00
Control 50	23.36±2.77	7.366±1.15	0.14±0.00	0.20±0.00	0.205±0.00	0.217±0.02
50 + P.F	18.10±0.63	18.70±0.00	0.23±0.04	0.28±0.00	0.246±0.00	0.286±0.01
Control 75	22.76±0.90	7.60±0.17	0.15±0.00	0.21±0.00	0.213±0.10	0.247±0.01
75+ P.F	16.76±2.71	17.76±0.15	0.24±0.00	0.29±0.00	0.254±0.00	0.306±0.01
Control 100	21.66±0.11	6.60±0.17	0.16±0.00	0.22±0.00	0.226±0.12	0.297±0.00
100+ P.F	15.76±3.23	16.96±0.76	0.25±0.00	0.29±0.00	0.277±0.00	0.337±0.00
Control 25	20.26±2.71	15.23±1.27	0.15±0.00	0.19±0.00	0.195±0.00	0.187±0.00
25+ P.A	24.83±0.57	21.93±0.46	0.22±0.01	0.25±0.00	0.237±0.01	0.287±0.00

Control 50	18.33±1.27	11.96±1.32	0.16±0.00	0.14±0.04	0.208±0.17	0.208±0.10
50+ P.A	23.83±2.48	19.36±0.51	0.23±0.01	0.27±0.00	0.248±0.00	0.283±0.00
Control 75	16.33±1.27	14.86±0.63	0.17±0.00	0.12±0.00	0.219±0.01	0.238±0.00
75+ P.A	20.26±2.71	17.43±0.63	0.24±0.01	0.27±0.00	0.258±0.00	0.308±0.15
Control 100	15.40±1.21	14.06±0.63	0.18±0.00	0.11±0.00	0.234±0.00	0.249±0.00
100 + P.A	19.76±3.23	16.96±0.76	0.24±0.17	0.29±0.00	0.268±0.01	0.336±0.16

*Individual readings n =3 for each plant were averaged and are presented with ±SD.

*Significant correlation was found between root length (RL), shoot length (SL), root, shoot dry weight (RDW, SDW) and root, shoot fresh weight (RFW, SFW) at 0.05 level of significance; P.F: Pseudomonas fluorescens; P.A: Pseudomonas aeruginosa.

Table 4.8.4: Biomass of *P. juliflora* plants grown in soil for 120 days in different F concentration viz. control, 25, 50, 75 and 100 mg kg^{-1} NaF soil after treatment with microbes P.F and P.A

Parameters(120 days)						
Treatment	**RL(cm)**	**SL(cm)**	**RDW(gm)**	**SDW(gm)**	**RFW (gm)**	**SFW (gm)**
Control	30.86±3.85	20.60±1.03	0.212±0.00	0.256±0.00	0.255±0.00	0.294±0.01
Control 25	26.40±0.69	16.66±3.08	0.167±0.00	0.187±0.00	0.191±0.00	0.218±0.03
25+ P.F	34.30±0.69	24.96±0.40	0.280±0.00	0.288±0.00	0.248±0.00	0.305±0.01
Control 50	24.83±5.31	14.63±2.13	0.180±0.00	0.189±0.00	0.218±0.01	0.237±0.00
50+ P.F	28.33±0.62	20.30±2.07	0.285±0.00	0.296±0.00	0.258±0.00	0.316±0.00
Control 75	23.33±3.92	14.66±1.61	0.185±0.01	0.192±0.00	0.235±0.00	0.258±0.00
75 + P.F	24.50±0.04	19.50±0.06	0.288±0.00	0.315±0.00	0.259±0.00	0.328±0.01
Control 100	12.76±0.01	10.43±0.23	0.203±0.01	0.195±0.00	0.247±0.00	0.269±0.00
100+ P.F	19.36±0.75	18.53±0.63	0.293±0.00	0.322±0.00	0.288±0.00	0.349±0.01
Control 25	26.33±2.54	14.76±1.09	0.165±0.00	0.196±0.00	0.210±0.02	0.210±0.02
25+ P.A	30.23±2.09	22.80±0.51	0.256±0.00	0.247±0.00	0.248±0.00	0.283±0.00
Control 50	21.73±1.67	12.76±0.05	0.181±0.00	0.206±0.01	0.229±0.01	0.221±0.01

50+ P.A	27.43±0.63	19.13±0.28	0.267±0.00	0.270±0.00	0.257±0.01	0.281±0.00
Control 75	19.06±1.09	11.86±1.15	0.188±0.00	0.221±0.00	0.244±0.01	0.237±0.00
75 + P.A	24.86±1.27	16.96±0.23	0.281±0.00	0.277±0.00	0.265±0.01	0.343±0.02
Control 100	17.26±0.40	10.60±0.11	0.204±0.00	0.260±0.05	0.248±0.00	0.259±0.00
100+ P.A	23.96±0.40	16.03±1.78	0.283±0.00	0.291±0.00	0.284±0.01	0.346±0.01

*Individual readings n =3 for each plant were averaged and are presented with ±SD.

*Significant correlation was found between root length (RL), shoot length (SL), root,

shoot dry weight (RDW, SDW) and root, shoot fresh weight (RFW, SFW) at 0.05 level of significance;

P.F: Pseudomonas fluorescens; P.A: Pseudomonas aeruginosa.

Table 4.9.1: Chlorophyll content of *P. juliflora* seedlings grown in soil for 30 and 60 days in different concentration of F viz. control, 25, 50, 75 and 100 mg kg^{-1} NaF soil after treatment with microbes P.F and P.A

Chlorophyll content (mg g^{-1}fw)								
30 days					60 days			
Treatment	**Chlorophyll a**	**Chlorophyll b**	**Total chlorophyll**	**Chlorophyll a/b**	**Chlorophyll a**	**Chlorophyll b**	**Total chlorophyll**	**Chlorophyll a/b**
Control	3.197±0.18	1.687±0.33	4.884±1.33	1.895	3.082±0.00	1.985±0.20	5.067±0.00	1.552
Control 25	0.597±0.00	0.325±0.18	0.922±0.09	1.835	0.803±0.04	0.185±0.00	0.988±0.09	4.340
25+ P.F	0.950±0.10	0.533±0.16	1.483±0.00	1.782	0.815±0.24	0.778±0.15	1.593±0.00	1.047
Control 50	0.593±0.00	0.327±0.00	0.920±0.19	1.812	0.807±0.00	0.410±0.06	1.217±0.04	1.968
50+ P.F	0.951±0.10	0.537±0.08	1.488±0.15	1.770	0.726±0.16	0.816±0.00	1.542±0.44	0.889
Control 75	0.587±0.07	0.338±0.07	0.925±0.14	1.736	0.811±0.18	0.417±0.05	1.228±0.00	1.944
75+P.F	1.027±0.37	0.513±0.15	1.540±0.16	2.001	0.836±0.00	0.859±0.16	1.695±0.00	0.973
Control 100	0.600±0.18	0.328±0.14	0.928±0.11	1.829	0.772±0.06	0.437±0.00	1.209±0.27	1.766
100+ P.F	1.026±0.00	0.506±0.17	1.532±0.18	2.024	0.819±0.06	0.874±0.07	1.693±0.00	0.937
Control 25	0.281±0.16	0.157±0.04	0.438±0.16	1.781	0.316±0.00	0.175±0.07	0.491±0.00	1.805
25+ P.A	0.954±0.16	0.499±0.11	1.453±0.18	1.910	1.187±0.07	0.640±0.00	1.827±0.28	1.854
Control 50	0.292±0.10	0.160±0.00	0.452±0.00	1.822	0.351±0.00	0.335±0.16	0.686±0.01	1.047
50+ P.A	0.961±0.00	0.486±0.16	1.447±0.01	1.975	0.816±0.00	0.783±0.12	1.599±0.07	1.042
Control 75	0.288±0.00	0.162±0.01	0.450±0.16	1.778	0.335±0.14	0.175±0.00	0.510±0.10	1.915
75+ P.A	1.041±0.10	0.458±0.00	1.499±0.00	2.269	0.844±0.06	0.856±0.16	1.700±0.00	0.985
Control 100	0.274±0.05	0.182±0.05	0.456±0.00	1.504	0.266±0.00	0.208±0.08	0.474±0.08	1.278
100 + P.A	1.046±0.00	0.455±0.00	1.501±0.00	2.298	0.888±0.20	0.930±0.00	1.818±0.05	0.954

*Individual readings n =3 for each plant were averaged and are presented with ±SD.

*Significant correlation was found between chlorophyll activities at 0.05 level of significance;

P.F: Pseudomonas fluorescens; P.A: Pseudomonas aeruginosa.

Table 4.9.2: Chlorophyll content of *P. juliflora* seedlings grown in soil for 90 and 120 days in different concentrations of F viz. control, 25, 50, 75 and 100 mg kg^{-1} NaF soil after treatment with microbes P.F and P.A

Chlorophyll content (mg g^{-1}fw)								
90 days					120 days			
Treatment	**Chlorophyll a**	**Chlorophyll b**	**Total chlorophyll**	**Chlorophyll a/b**	**Chlorophyll a**	**Chlorophyll b**	**Total chlorophyll**	**Chlorophyll a/b**
Control	4.320±0.50	2.183±0.55	6.503±0.55	1.978	4.090±0.50	1.860±0.30	5.951±0.55	2.198
Control 25	1.050±0.00	0.284±0.07	1.334±0.00	3.697	0.968±0.04	0.139±0.04	1.107±0.55	6.964
25+ P.F	2.625±0.44	0.858±0.07	3.484±0.07	3.060	1.931±0.00	1.623±0.30	3.554±0.04	1.189
Control 50	1.097±0.50	0.379±0.00	1.476±0.00	2.894	1.020±0.40	0.353±0.05	1.373±0.04	2.889
50+ P.F	2.639±0.00	0.856±0.08	3.495±0.07	3.082	1.945±0.05	1.623±0.05	3.568±0.00	1.198
Control 75	1.032±0.00	0.396±0.05	1.428±0.07	2.606	1.031±0.00	0.365±0.02	1.396±0.33	2.821
75+P.F	2.671±0.50	0.977±0.05	3.648±0.09	2.733	2.035±0.04	1.565±0.04	3.600±0.03	1.300
Control 100	0.959±0.09	0.414±0.09	1.373±0.05	2.316	0.970±0.06	0.391±0.10	1.361±0.03	2.480
100+ P.F	3.527±0.55	0.657±0.05	4.184±0.09	5.368	3.329±0.04	1.101±0.54	4.430±0.02	3.023
Control 25	0.448±0.00	0.232±0.09	0.681±0.06	1.931	0.444±0.00	0.152±0.03	0.596±0.06	2.921
25+ P.A	2.526±0.44	0.165±0.06	2.691±0.45	15.30	0.546±0.13	0.674±0.02	1.220±0.04	0.810
Control 50	0.447±0.00	0.232±0.05	0.679±0.07	1.926	0.437±0.05	0.176±0.10	0.613±0.05	2.482
50+ P.A	2.440±0.60	0.937±0.05	3.377±0.50	2.604	2.377±0.04	1.480±0.20	3.857±0.02	1.606
Control 75	0.431±0.00	0.189±0.09	0.620±0.07	2.280	0.406±0.04	0.158±0.05	0.564±0.00	2.569
75+ P.A	2.937±0.10	0.893±0.04	3.830±0.00	3.288	2.892±0.03	1.277±0.04	4.169±0.00	2.264
Control 100	0.305±0.06	0.220±0.07	0.525±0.10	1.386	0.294±0.04	0.202±0.04	0.496±0.01	1.455
100+ P.A	3.352±0.16	0.909±0.00	4.261±0.68	3.687	2.476±0.04	1.366±0.04	3.842±0.02	1.812

*Individual readings n =3 for each plant were averaged and are presented with ±SD.

*Significant correlation was found between chlorophyll activities at 0.05 level of significance;

P.F: Pseudomonas fluorescens; P.A: Pseudomonas aeruginosa.

Table 4.10.1: Enzymatic activity of *P. juliflora* plants grown in soil for 30 and 60 days in different F concentrations viz. control, 25, 50, 75 and 100 mg kg^{-1} NaF soil after treatment with microbes P.F and P.A

Antioxidant enzyme activity (µg g^{-1}fw)						
30 days				**60 days**		
Treat-ment	**Peroxidase**	**Catalase**	**Superoxi-dase**	**Peroxidase**	**Catalase**	**Superoxidase**
Control	0.017±0.04	0.085±0.00	0.017±0.00	0.024±0.00	0.121±0.00	0.125±0.01
Control 25	0.149±0.05	0.114±0.00	0.112±0.00	0.150±0.00	0.124±0.02	0.161±0.05
25+ P.F	0.124±0.00	0.222±0.01	0.229±0.00	0.135±0.00	0.274±0.03	0.245±0.00
Control 50	0.153±0.00	0.133±0.01	0.130±0.01	0.159±0.00	0.136±0.02	0.232±0.01
50+ P.F	0.125±0.00	0.245±0.00	0.234±0.00	0.180±0.00	0.213±0.00	0.258±0.03
Control 75	0.155±0.00	0.137±0.00	0.135±0.00	0.169±0.01	0.141±0.02	0.240±0.01
75+ P.F	0.256±0.01	0.270±0.00	0.240±0.01	0.225±0.00	0.238±0.00	0.329±0.01
Control 100	0.266±0.00	0.181±0.00	0.155±0.00	0.177±0.01	0.150±0.03	0.266±0.01
100+ P.F	0.275±0.01	0.273±0.02	0.259±0.01	0.226±0.00	0.329±0.01	0.367±0.02
Control 25	0.106±0.00	0.118±0.00	0.125±0.01	0.128±0.00	0.131±0.02	0.258±0.02
25+ P.A	0.115±0.00	0.212±0.00	0.212±0.00	0.135±0.00	0.141±0.04	0.260±0.00
Control 50	0.112±0.00	0.138±0.00	0.133±0.01	0.134±0.00	0.135±0.02	0.272±0.00
50 + P.A	0.213±0.00	0.223±0.00	0.233±0.00	0.136±0.00	0.151±0.08	0.290±0.00
Control 75	0.115±0.01	0.138±0.00	0.142±0.00	0.143±0.00	0.138±0.03	0.274±0.03
75+ P.A	0.221±0.00	0.231±0.01	0.245±0.00	0.156±0.01	0.159±0.01	0.275±0.00
Control 100	0.121±0.00	0.158±0.00	0.156±0.00	0.180±0.01	0.149±0.04	0.279±0.04
100+ P.A	0.227±0.00	0.239±0.00	0.265±0.17	0.186±0.01	0.217±0.01	0.280±0.02

*Individual readings n =3 for each plant were averaged and are presented with ±SD.

*Significant correlation was found between peroxidase, catalase and superoxidase activity at 0.05 level of significance; P.F: Pseudomonas fluorescens; P.A: Pseudomonas aeruginosa.

Table 4.10.2: Enzymatic activity of *P. juliflora* plants grown in soil for 90 and 120 days in different F concentrations viz. control, 25, 50, 75 and 100 mg kg^{-1} NaF soil after treatment with microbes P.F and P.A

Antioxidant enzyme activity (μg g^{-1}fw)						
90 days				**120 days**		
Treatment	**Peroxidase**	**Catalase**	**Superoxi-dase**	**Peroxi-dase**	**Catalase**	**Superoxi-dase**
Control	0.056±0.00	0.134±0.00	0.148±0.01	0.100±0.00	0.103±0.00	0.116±0.00
Control 25	0.166±0.00	0.179±0.08	0.265±0.10	0.120±0.01	0.150±0.04	0.122±0.01
25 + P.F	0.173±0.02	0.418±0.00	0.363±0.08	0.132±0.00	0.160±0.00	0.132±0.00
Control 50	0.129±0.00	0.193±0.02	0.333±0.00	0.140±0.00	0.160±0.01	0.140±0.00
50+ P.F	0.181±0.01	0.436±0.16	0.483±0.17	0.159±0.00	0.165±0.00	0.146±0.01
Control 75	0.145±0.00	0.212±0.05	0.332±0.01	0.176±0.01	0.166±0.00	0.153±0.00
75+ P.F	0.219±0.00	0.463±0.05	0.508±0.14	0.179±0.00	0.170±0.02	0.158±0.01
Control 100	0.154±0.00	0.304±0.05	0.398±0.05	0.183±0.02	0.175±0.00	0.162±0.00
100+ P.F	0.218±0.00	0.479±0.01	0.692±0.08	0.214±0.00	0.180±0.03	0.170±0.01
Control 25	0.135±0.00	0.245±0.12	0.270±0.02	0.115±0.00	0.141±0.00	0.113±0.00
25+ P.A	0.145±0.00	0.262±0.04	0.295±0.10	0.114±0.00	0.128±0.00	0.124±0.01
Control 50	0.140±0.00	0.265±0.04	0.280±0.03	0.135±0.00	0.150±0.00	0.124±0.01
50+ P.A	0.217±0.00	0.323±0.17	0.359±0.08	0.118±0.01	0.136±0.00	0.128±0.00
Control 75	0.148±0.01	0.311±0.11	0.284±0.00	0.155±0.00	0.163±0.02	0.128±0.01
75+ P.A	0.228±0.00	0.380±0.05	0.389±0.06	0.129±0.01	0.145±0.01	0.140±0.01
Control 100	0.145±0.01	0.378±0.23	0.290±0.00	0.170±0.00	0.149±0.04	0.141±0.00
100+ P.A	0.238±0.01	0.401±0.11	0.587±0.08	0.136±0.02	0.150±0.02	0.150±0.01

*Individual readings n =3 for each plant were averaged and are presented with ±SD.

*Significant correlation was found between peroxidase, catalase and superoxidase activity at 0.05 level of significance; P.F: Pseudomonas fluorescens; P.A: Pseudomonas aeruginosa.

Table 4.11: Fluoride accumulation in root and shoot of *P. juliflora* seedlings were grown in soil for 30, 60, 90 and 120 days in different concentrations of viz. control, 25, 50, 75 and 100 mg kg^{-1} NaF after treatment with microbes P.F and P.A

Fluoride content (mg kg^{-1})								
Root uptake					**Shoot uptake**			
Treatment	**30 days**	**60 days**	**90 days**	**120 days**	**30 days**	**60 days**	**90 days**	**120 days**
Control	0.11±0.01	0.11±0.00	0.15±0.00	1.06±0.00	0.04±0.00	0.07±0.00	0.08±0.00	0.78±0.02
Control 25	1.02±0.06	1.13±0.15	1.16±0.20	5.59±0.75	0.33±0.00	0.42±0.01	0.56±0.00	3.69±0.43
25+ P.F	1.16±0.28	1.30±0.26	2.03±0.05	7.86±0.23	0.89±0.00	1.10±0.00	1.30±0.10	5.67±1.15
Control 50	0.91±0.01	1.38±0.18	1.73±0.15	18.25±0.44	0.65±0.00	0.74±0.06	0.88±0.10	12.06±0.03
50+ P.F	3.20±0.26	3.30±0.26	4.13±0.1[illegible]	21.11±1.02	1.60±0.04	2.50±0.07	3.03±0.05	19.34±0.27
Control 75	1.25±0.06	3.20±0.26	3.33±0.15	21.66±7.9	0.86±0.01	0.94±0.03	1.17±0.04	14.01±0.00
75+ P.F	3.66±0.15	4.46±0.50	7.20±0.26	35.92±0.78	2.51±0.05	3.43±0.20	4.10±0.10	30.00±0.25
Control 100	3.18±0.06	3.30±0.10	3.40±0.36	22.62±0.54	1.29±0.05	1.60±0.10	2.70±0.10	24.27±0.00
100+ P.F	13.36±0.25	18.23±0.40	21.53±0.46	40.55±0.31	6.51±0.05	8.75±0.15	12.23±0.20	44.18±0.26
Control 25	1.12±0.06	1.16±0.20	1.10±0.10	4.42±0.01	0.33±0.01	0.42±0.01	0.61±0.00	2.29±0.01
25+ P.A	1.16±0.15	1.40±0.30	1.23±0.25	5.23±0.73	0.55±0.01	0.74±0.01	0.80±0.01	4.78±0.31
Control 50	1.30±0.10	1.47±0.10	1.56±0.05	14.16±0.15	0.53±0.02	0.64±0.02	1.05±0.01	13.13±0.00
50+ P.A	3.10±0.10	3.36±0.11	4.06±0.11	16.55±1.52	0.86±0.01	0.95±0.05	1.10±0.10	14.16±0.22
Control 75	2.40±0.10	3.13±0.23	3.23±0.05	20.25±0.25	0.86±0.01	0.90±0.05	1.26±0.06	14.49±0.02
75 + P.A	3.36±0.25	3.76±0.11	5.26±0.30	21.45±0.63	1.10±0.12	1.52±0.02	1.70±0.10	15.33±0.12
Control 100	2.73±0.05	3.26±0.23	3.66±0.15	18.29±0.09	1.65±0.01	2.30±0.43	3.10±0.00	16.33±0.57
100+ P.A	12.56±0.20	14.60±0.30	17.23±0.40	35.56±0.63	6.41±0.00	7.53±0.30	9.86±0.75	37.00±0.20

*Individual readings n =3 for each plant were averaged and are presented with ±SD.

*Significant correlation was found between Fluoride content at 0.05 level of significance;

P.F: Pseudomonas fluorescens; P.A: Pseudomonas aeruginosa.

Table 4.12: Remaining F concentration in soil after harvesting the plantlets of *P. juliflora* for 30, 60, 90 and 120 days in different concentration of F viz. control, 25, 50, 75, and 100 mg kg^{-1} NaF after treatment with microbes P.F and P.A

Fluoride content (mg kg^{-1})				
Treatment	**30 days**	**60 days**	**90 days**	**120 days**
Control	2.41±0.30	2.14±0.00	2.00±0.70	1.56±0.40
Control 25	23.63±0.00	23.41±1.00	23.21±2.00	15.72±0.00
25 + P.F	22.93±0.30	22.60±1.00	21.66±2.00	11.47±3.00
Control 50	48.43±0.00	47.86±0.90	47.21±2.00	19.69±2.00
50+ P.F	45.19±0.30	44.18±1.00	42.83±0.90	9.55±0.00
Control 75	72.88±0.20	70.85±0.00	70.40±3.00	39.33±0.00
75+ P.F	68.81±1.00	67.10±0.80	63.70±0.00	9.08±1.00
Control 100	95.52±2.00	95.10±2.00	93.50±4.00	53.11±0.50
100+ P.F	80.11±5.00	73.01±0.00	66.23±0.00	15.27±0.00
Control 25	23.56±0.00	23.41±2.00	23.33±5.00	18.29±0.50
25+ P.A	23.27±6.00	22.85±4.00	22.95±0.60	14.99±0.10
Control 50	48.16±0.50	47.95±0.60	47.54±3.00	22.71±0.70
50+ P.A	46.03±0.30	45.67±0.00	44.83±0.00	19.29±0.50
Control 75	71.73±0.90	70.95±1.00	70.59±2.00	36.26±0.67
75+ P.A	70.52±0.30	69.70±0.00	68.03±0.80	38.22±0.70
Control 100	95.61±0.40	94.40±1.00	93.63±3.00	65.38±0.40
100+ P.A	81.02±0.30	77.86±1.00	72.90±1.00	27.44±0.70

*Significant correlation was found between Fluoride content at 0.05 level of significance; P.F: Pseudomonas fluorescens; P.A: Pseudomonas aeruginosa.

Table 4.13: Translocation and bioaccumulation factor values (shoot/soil concentration ratio) of *P. juliflora* plants grown in soil under different concentration of F viz. control, 25, 50, 75 and 100 mg kg^{-1} NaF soil for 30, 60, 90 and 120 days after treatment with microbes P.F and P.A

Fluoride content								
Translocation					**Bioaccumulation**			
Treatment	**30 days**	**60 days**	**90 days**	**120 days**	**30 days**	**60 days**	**90 days**	**120 days**
Control	0.39	0.64	0.54	0.73	0.08	0.22	0.38	0.50
Control 25	0.32	0.39	0.52	0.66	0.01	0.01	0.02	0.23
25 + P.F	0.76	0.84	0.63	0.72	0.03	0.04	0.05	0.49
Control 50	0.70	0.53	0.60	0.66	0.01	0.01	0.02	0.61
50+ P.F	0.50	0.76	0.73	0.91	0.03	0.05	0.07	2.02
Control 75	0.69	0.29	0.38	0.64	0.01	0.01	0.01	0.35
75+ P.F	0.68	0.76	0.56	0.83	0.03	0.05	0.06	3.30
Control 100	0.40	0.48	0.91	1.07	0.01	0.01	0.03	0.45
100+ P.F	0.48	0.48	0.56	1.08	0.08	0.11	0.18	2.89
Control 25	0.30	0.36	0.51	0.51	0.01	0.01	0.02	0.12
25+ P.A	0.47	0.53	0.65	0.91	0.02	0.03	0.03	0.31
Control 50	0.41	0.45	0.56	0.92	0.01	0.01	0.01	0.57
50 + P.A	0.28	0.28	0.27	0.85	0.01	0.02	0.02	0.73
Control 75	0.35	0.29	0.36	0.71	0.01	0.01	0.01	0.39
75+ P.A	0.32	0.40	0.32	0.71	0.01	0.02	0.02	0.40
Control 100	0.60	0.70	0.73	0.89	0.01	0.02	0.02	0.24
100+ P.A	0.51	0.51	0.57	1.04	0.79	0.09	0.13	1.34

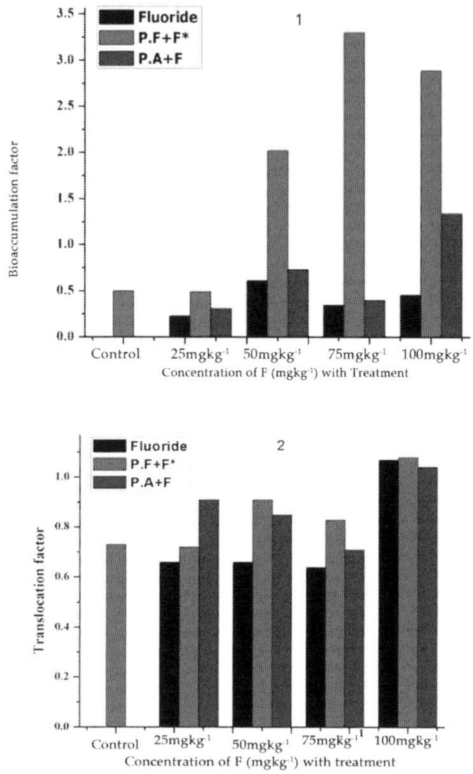

Figure 4.14: Effect of the microbial consortium on bioaccumulation factor and translocation factor under F stress for 120 days (1-bioaccumulation factor and 2-translocation factor, P.F- *Pseudomonas fluorescens* and P.A- *Pseudomonas aeruginosa* where significant value*).

4.11 EFFECT OF CHELATES (EDTA AND CA) ON GROWTH PARAMETERS

The germination percentage was 70 percent in the case of EDTA, 60 percent in CA and 50 percent at 100 mg kg^{-1} NaF with combined EDTA and CA which is shown in (Table 4.14). The effect of EDTA, CA and combined EDTA with CA on plant growth was observed (Figure 4.15 and Table 4.15) and the length was measured under different treatments which are shown in (Figure 4.15). Treated plants grew significantly better than control plants. The plant growth was analyzed in all the parameters such as RL, SL, RFW, SFW, RDW and SDW. Without additional EDTA, CA and combined EDTA with CA amendment, plants have not shown promoting effects on the root, shoot length and biomass production of *P. juliflora*. However, the length and biomass of plant tissues were significantly ($p \leq 0.05$) affected by 100 mg kg^{-1}NaF. However, there was significantly ($p \leq 0.05$) increased both the shoot length (30.26, 27.50 and 26.00 cm) and the shoot biomass (FW from 0.428, 0.357 and 0.285 g and DW from 0.381, 0.314 and 0.280 g) in EDTA as compared to CA and combined EDTA with CA. Meanwhile, root length was observed (44.56, 38.70 and 32.70 cm) and the root biomass production (FW from 0.317, 0.313 and 0.260 g, and DW from 0.385, 0.277 and 0.235 g). Excess amounts of cobalt, chromium and copper had a negative effect on the biomass of cauliflower (Chatterjee and Chatterjee, 2000).Such a reduction in the growth of Cd stressed plants could be ascribed mainly to inhibited cell division and/or cell enlargement (Davies *et al*., 1991). Our results have shown that the increase in the F concentration decreased the height of plants and their biomass. The same results were found in bean plant reduction in total biomass and dry weight of seedlings with increasing Cd concentration (Bhardwaj *et al*.,2009). However treated plant with EDTA, CA and combined EDTA with CA have significantly increased their biomass.

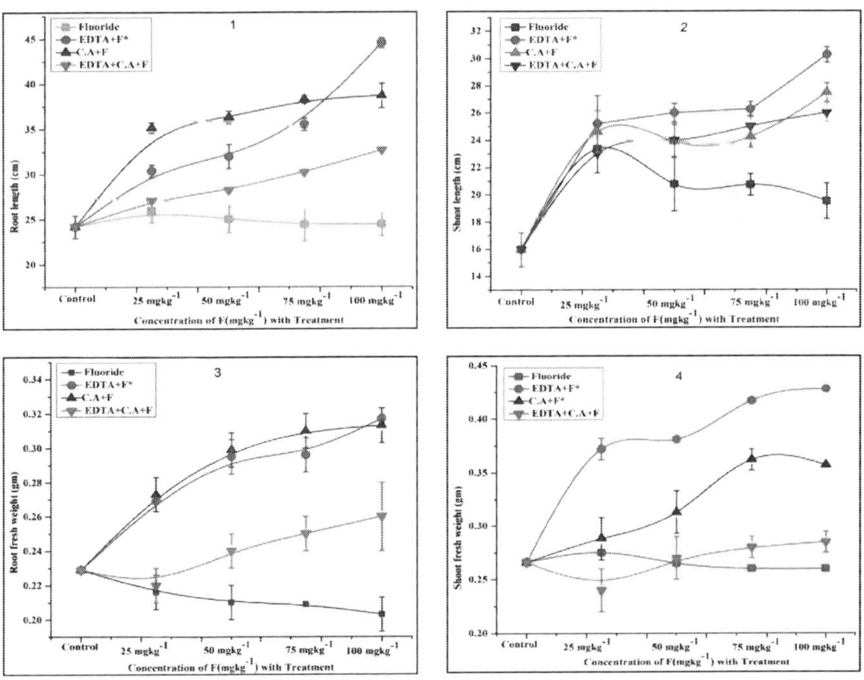

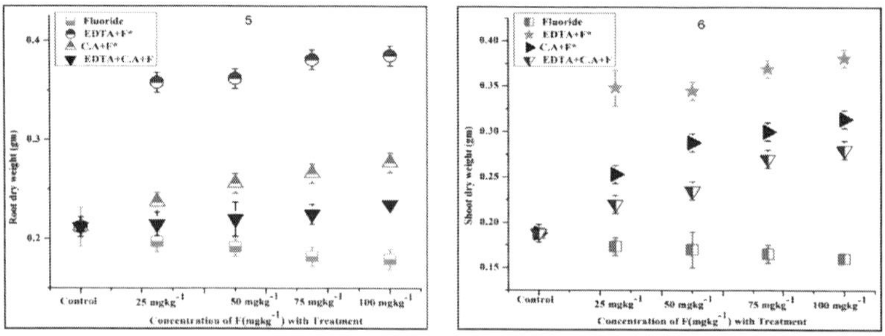

Figure 4.15: Effect of chelates (EDTA and CA) on plant growth and biomass under F-contaminated soil for 120 days (1-root length, 2-shoot length,3-root dry weight, 4-shoot dry weight, 5-root fresh weight, 6-shoot fresh weight and EDTA-Ethylene diamine tetraacetic acid and CA-Citric acid significant value*).

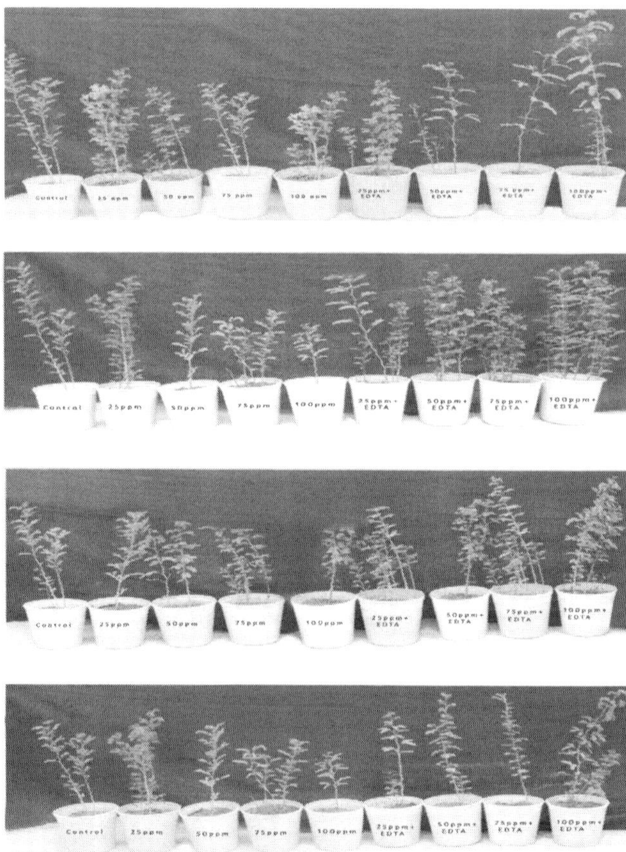

(1)

(2)

(3)

(4)

Plate 4.15.A: Effect of ethylene diamine tetraacetic acid (EDTA) on plant growth for 120 days of pot experiment with different concentrations of NaF viz. control, 25, 50, 75 and 100 mg kg^{-1} soil.

(1) 5 mM EDTA

(2) 10 mM EDTA

(3) 15 mM EDTA

(4) 20 mM EDTA

(1)

(2)

(3)

(4)

Plate 4.15.B: Effect of citric acid (CA) on plant growth for 120 days of pot experimentwith different concentrations of NaF viz. control, 25, 50, 75 and 100 mg kg^{-1} soil.

(1) 5 mM CA

(2) 10 mM CA

(3) 15 mM CA

(4) 20 mM CA

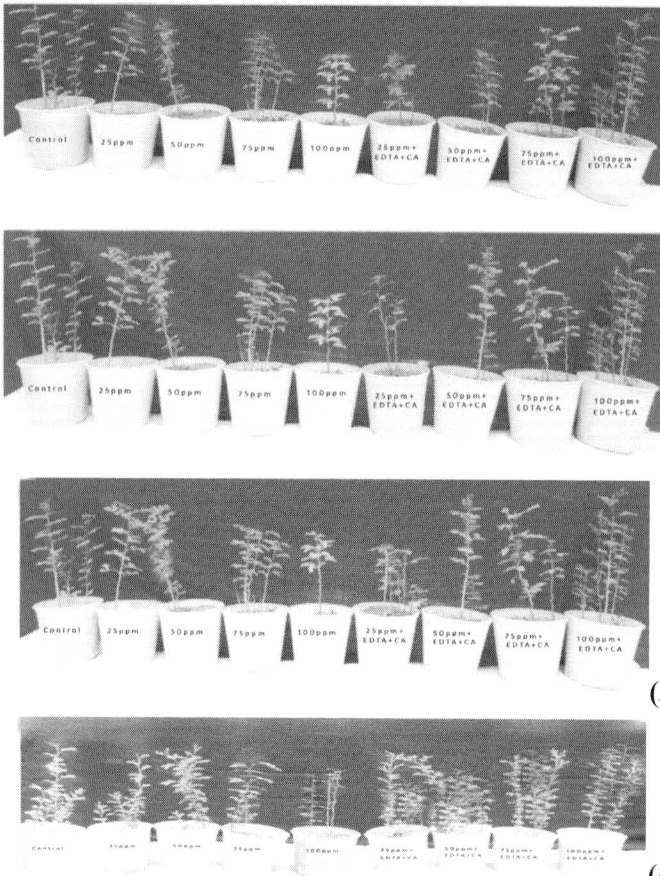

(1)

(2)

(3)

(4)

Plate 4.15.C: Effect of ethylene diamine tetraacetic acid (EDTA) and citric acid (CA) on plant growth for 120 days of pot experiment with different concentrations of NaF viz. control, 25, 50, 75 and 100 mg kg^{-1} soil.

(1) 5 mM EDTA+CA
(2) 10 mM EDTA+CA
(3) 15 mM EDTA+CA
(4) 20 mM EDTA+CA

4.12 EFFECT OF CHELATES (EDTA AND CA) ON CHLOROPHYLL CONTENTS

In the present study, F toxicity was not found at 15 days after the given treatment by EDTA, C.A and combined EDTA and CA. There were no external symptoms seen on leaves of plants treated with 25, 50, 75 and 100 mg kg-1 NaF. Total Chl was gradually decreased with increasing concentration of F after 30 days or upto 90 days. In leaves, chlorophyll contents were decreased to Chl a (1.001) Chl b (0.889), total Chl (1.890) Chl a/b (1.125) at 100 mg kg-1 in soil.

The treated leaves with EDTA were found to have significantly increased value Chl a (1.914) Chl b (0.646), total Chl (2.560), Chl a/b (2.962) of chlorophyll as compared to separate results with CA and combined results of EDTA with CA (Figure s 4.16 and Table 4.17).

Thus to summarise the present study, it can be said that overall there is a decrease in the total Chl content with an increase in F concentration was found after 30 days of F exposure. As one of the most important plant organs, leaves play a role in capturing light and making food via the photosynthesis process. It is one of the most sensitive processes. There is enormous change observed after 90 days that total chl increased up to 120 days and gradually protected pigment from F stress after given treatment with chelates EDTA, CA and combined EDTA+CA.

The decline in chlorophyll content in plants exposed to F is believed to be due to: (a) inhibition of enzymes associated with chlorophyll biosynthesis(John *et al.*, 2009), (b) inhibition of uptake and transportation of other metal elements such as Fe, Mn and Zn by antagonistic effects (Jayakumar *et al.*, 2009).

It is likely that the higher concentration of F in our treatments, which is 100 mg kg^{-1}, resulted in no visible injury. It was below the threshold level required for such an effect. Thus, very high doses of soluble F in the root would be needed to cause harmful effects. Previous findings have also confirmed that F causes a decrease inthe chlorophyll content in *Pisum sativum*, *Oryza sativa*and*Triticum aestivum* (Gupta *et al.*,2009; Bhargava and Bhardwaj 2010; Saini *et al.*, 2012). Overall, our result also found that exposure time is more essential than concentration of F accumulations in the leaves and inhibition of pigment synthesis.

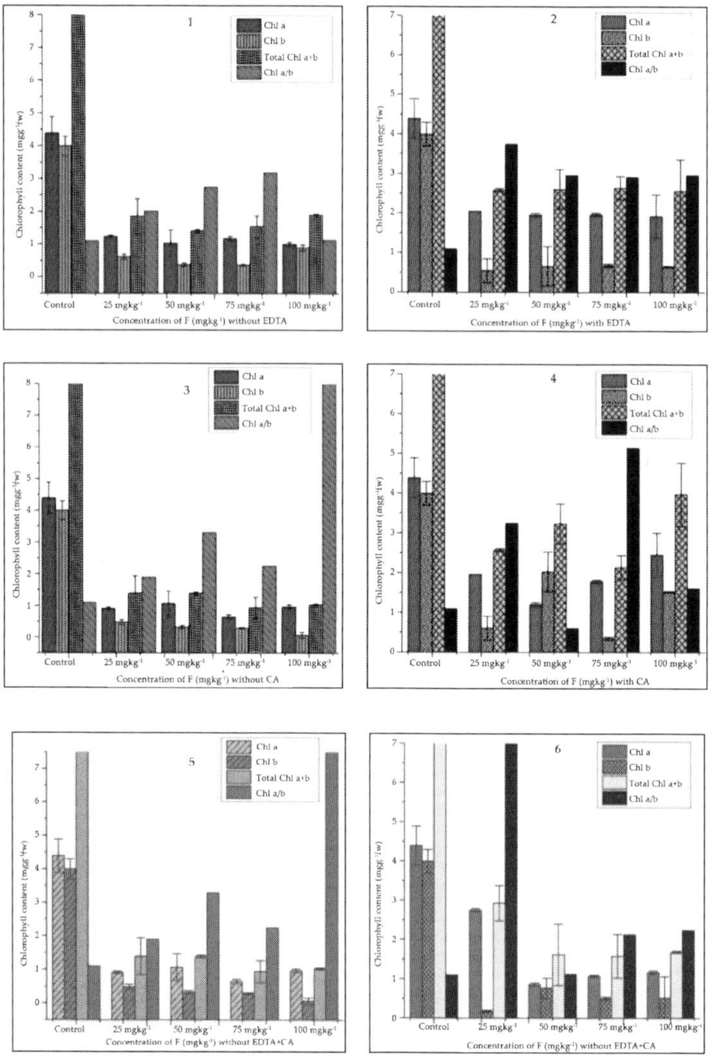

Figure 4.16: Effect of chelates (EDTA and CA) on chlorophyll contents of *P. juliflora* under F- stress for 120 days (1,2-EDTA- Ethylene diamine tetraacetic acid, 3,4-CA and 5,6-Ethylene diamine tetraacetic acid +Citric acid where significant value*).

Table 4.14: Germination percentages of *P. juliflora* seedlings grown in the soil after 10 days in different concentrations of F viz. control, 25, 50, 75 and 100 mg kg^{-1} NaF soil after treatment with chelates EDTA and CA.

Germination content %				
Treatment %	**5 mM Kg^{-1}**	**10 mMKg^{-1}**	**15 mM Kg^{-1}**	**20 mM Kg^{-1}**
Control	60%	60%	65%	55%
Control 25	40%	45%	40%	40%
25+ EDTA	75%	70%	75%	70%
Control 50	55%	35%	50%	45%
50+ EDTA	45%	65%	60%	70%
Control 75	45%	45%	35%	30%
75 + EDTA	70%	75%	65%	60%
Control 100	10%	10%	20%	20%
100+ EDTA	65%	65%	65%	70%
Control 25	55%	65%	60%	50%
25+ C.A	55%	60%	65%	60%
Control 50	50%	45%	45%	45%
50+ C.A	60%	60%	60%	75%
Control 75	40%	35%	30%	40%
75+ C.A	65%	65%	60%	55%
Control 100	15%	20%	20%	20%
100+ C.A	60%	65%	60%	60%
Control 25	75%	70%	75%	70%
25+ EDTA and C.A	55%	35%	50%	45%
Control 50	45%	65%	60%	70%
50+ EDTA and C.A	45%	45%	35%	30%

Control 75	70%	75%	65%	60%
75+ EDTA and C.A	40%	40%	30%	20%
Control 100	45%	60%	65%	20%
100+ EDTA and C.A	55%	55%	50%	50%

*EDTA: Ethylene diamine tetraacetic acid; C.A: Citric acid.

Table 4.15.1: Biomass of *P. juliflora* plants grown in soil for 30 days in different F concentration viz. control, 25, 50, 75 and 100 mg kg^{-1} NaF soil after treatment with chelates EDTA and CA

Concentration (5 mM)						
Treatment	**RL(cm)**	**SL(cm)**	**RDW(gm)**	**SDW(gm)**	**RFW (gm)**	**SFW (gm)**
Control	10.03±0.20	7.33±0.05	0.109±0.00	0.232±0.03	0.126±0.00	0.202±0.00
Control 25	5.53±1.17	6.06±0.11	0.066±0.00	0.092±0.08	0.114±0.00	0.132±0.00
25 + EDTA	9.13±0.15	7.43±0.45	0.127±0.01	0.262±0.06	0.180±0.00	0.298±0.01
Control 50	7.10±0.26	4.13±0.16	0.084±0.00	0.101±0.00	0.130±0.01	0.143±0.00
50+ EDTA	10.87±0.72	5.36±0.23	0.114±0.00	0.272±0.06	0.187±0.00	0.306±0.01
Control 75	9.83±0.33	3.16±0.20	0.042±0.00	0.112±0.09	0.125±0.08	0.166±0.00
75+ EDTA	12.06±0.55	4.30±0.43	0.153±0.00	0.243±0.00	0.194±0.00	0.312±0.00
Control 100	12.10±0.1	3.11±0.10	0.065±0.00	0.118±0.00	0.120±0.04	0.165±0.00
100 + EDTA	17.22±0.10	4.16±0.20	0.131±0.00	0.273±0.00	0.181±0.08	0.328±0.01
Control 25	6.23±0.25	4.56±2.34	0.084±0.00	0.094±0.08	0.100±0.08	0.127±0.00
25+ C.A	9.20±0.26	5.13±0.15	0.121±0.02	0.170±0.00	0.166±0.00	0.184±0.01
Control 50	10.83±0.50	3.45±0.22	0.077±0.00	0.155±0.00	0.117±0.00	0.148±0.00
50+ C.A	14.31±0.36	4.10±0.10	0.129±0.07	0.228±0.01	0.175±0.05	0.238±0.00
Control 75	11.20±0.20	3.54±0.21	0.073±0.00	0.142±0.00	0.126±0.00	0.187±0.00
75+ C.A	15.30±0.43	4.53±0.40	0.135±0.00	0.243±0.00	0.181±0.07	0.248±0.00
Control 100	12.50±0.43	3.21±0.00	0.077±0.00	0.141±0.00	0.137±0.02	0.203±0.00
100 + C.A	16.30±0.43	4.30±0.43	0.133±0.00	0.274±0.00	0.184±0.07	0.255±0.00
Control 25	6.23±0.25	4.56±2.34	0.084±0.00	0.094±0.08	0.100±0.08	0.127±0.00
25 + EDTA and C.A	7.20±0.26	5.00±0.00	0.111±0.02	0.110±0.00	0.106±0.00	0.130±0.01
Control 50	10.83±0.51	3.45±0.22	0.077±0.00	0.155±0.00	0.117±0.00	0.148±0.00
50+ EDTA and C.A	11.37±0.36	4.00±0.00	0.115±0.07	0.118±0.01	0.118±0.05	0.145±0.00
Control 75	11.20±0.20	3.54±0.21	0.073±0.00	0.142±0.00	0.122±0.00	0.187±0.00

75+ EDTA and C.A	12.30±0.43	4.00±0.00	0.125±0.00	0.123±0.00	0.125±0.07	0.150±0.00
Control 100	12.50±0.43	3.21±0.00	0.077±0.00	0.141±0.00	0.137±0.02	0.203±0.00
100+ EDTA and C.A	13.30±0.43	4.00±0.00	0.130±0.00	0.135±0.00	0.140±0.07	0.155±0.00

Concentration (10 mM)						
Treatment	**RL(cm)**	**SL(cm)**	**RDW(gm)**	**SDW(gm)**	**RFW (gm)**	**SFW (gm)**
Control	10.03±0.20	7.33±0.05	0.109±0.00	0.232±0.03	0.126±0.00	0.202±0.00
Control 25	5.76±0.65	6.33±0.32	0.64±0.00	0.145±0.00	0.124±0.00	0.138±0.00
25+ EDTA	7.00±0.08	6.46±0.32	0.123±0.02	0.265±0.06	0.185±0.00	0.303±0.00
Control 50	12.16±0.40	6.11±0.22	0.090±0.00	0.134±0.00	0.127±0.00	0.143±0.00
50+ EDTA	15.90±0.62	8.16±0.20	0.123±0.00	0.276±0.06	0.195±0.00	0.317±0.00
Control 75	16.50±0.17	6.20±0.10	0.063±0.00	0.132±0.00	0.132±0.00	0.169±0.00
75+ EDTA	21.13±0.41	4.33±0.32	0.144±0.00	0.244±0.01	0.204±0.00	0.318±0.08
Control 100	17.60±0.95	4.33±0.15	0.072±0.00	0.124±0.02	0.131±0.01	0.180±0.00
100+ EDTA	18.40±0.20	5.40±0.43	0.143±0.00	0.281±0.00	0.191±0.00	0.329±0.00
Control 25	8.68±0.003	3.54±0.00	0.073±0.00	0.156±0.00	0.114±0.00	0.137±0.00
25 + C.A	10.83±0.76	4.2±0.26	0.115±0.00	0.181±0.01	0.177±0.00	0.215±0.00
Control 50	11.43±0.70	4.22±0.02	0.085±0.00	0.145±0.09	0.122±0.00	0.150±0.01
50+ C.A	12.33±0.25	6.53±0.40	0.135±0.00	0.243±0.00	0.189±0.08	0.242±0.00
Control 75	11.06±0.06	5.12±0.33	0.083±0.00	0.145±0.00	0.135±0.00	0.191±0.00
75 + C.A	14.03±0.90	7.16±0.11	0.142±0.00	0.252±0.00	0.193±0.01	0.253±0.00
Control 100	12.90±0.26	6.19±0.06	0,084±0.02	0.153±0.09	0.141±0.00	0.195±0.00
100 + C.A	15.43±0.25	7.26±0.20	0.143±0.00	0.277±0.01	0.189±0.04	0.276±0.00
Control 25	8.68±0.003	3.54±0.00	0.073±0.00	0.156±0.00	0.114±0.00	0.137±0.00
25+ EDTA and C.A	9.00±0.76	4.0±0.26	0.105±0.00	0.160±0.01	0.160±0.00	0.205±0.00
Control 50	11.43±0.70	4.22±0.02	0.085±0.00	0.145±0.09	0.122±0.00	0.150±0.01
50+ EDTA and C.A	10.00±0.25	4.60±0.40	0.115±0.00	0.165±0.00	0.165±0.08	0.215±0.00
Control 75	11.06±0.06	5.12±0.33	0.083±0.00	0.145±0.00	0.135±0.00	0.191±0.00
75+ EDTA and C.A	11.00±0.90	5.50±0.11	0.120±0.00	0.170±0.00	0.170±0.01	0.220±0.00
Control 100	12.90±0.26	6.19±0.06	0.084±0.02	0.153±0.09	0.141±0.00	0.195±0.00
100+ EDTA and C.A	13.00±0.25	6.70±0.20	0.125±0.00	0.175±0.01	0.175±0.04	0.225±0.00

Concentration (15 mM)						
Treatment	**RL(cm)**	**SL(cm)**	**RDW(gm)**	**SDW(gm)**	**RFW (gm)**	**SFW (gm)**
Control	10.03±0.20	7.33±0.05	0.109±0.00	0.232±0.03	0.126±0.00	0.202±0.00
Control 25	7.89±0.17	5.33±0.23	0.073±0.00	0.084±0.00	0.136±0.00	0.143±0.07

25 + EDTA	8.10±0.10	6.53±0.23	0.126±0.00	0.285±0.06	0.200±0.01	0.314±0.00
Control 50	18.3±0.17	6.11±0.33	0.093±0.00	0.144±0.00	0.147±0.03	0.158±0.00
50+ EDTA	23.16±0.05	7.10±0.10	0.134±0.00	0.248±0.17	0.203±0.00	0.320±0.00
Control 75	19.06±0.16	8.89±0.54	0.075±0.00	0.135±0.00	0.144±0.00	0.173±0.00
75 + EDTA	19.40±0.36	9.23±0.23	0.152±0.00	0.266±0.00	0.235±0.00	0.325±0.00
Control 100	20.56±0.56	9.34±0.12	0.067±0.01	0.129±0.00	0.152±0.01	0.189±0.00
100 + EDTA	18.20±0.34	12.26±0.2	0.131±0.00	0.284±0.03	0.201±0.00	0.341±0.00
Control 25	10.17±1.81	4.13±0.55	0.089±0.00	0.125±0.03	0.132±0.00	0.147±0.00
25+ C.A	11.26±0.20	6.13±0.05	0.128±0.01	0.190±0.00	0.183±0.00	0.215±0.00
Control 50	12.13±0.96	4.44±0.33	0.092±0.00	0.134±0.08	0.136±0.00	0.151±0.01
50+ C.A	15.70±0.72	4.93±0.47	0.140±0.00	0.235±0.00	0.190±0.00	0.242±0.00
Control 75	15.56±2.67	5.21±0.33	0.098±0.00	0.129±0.03	0.143±0.00	0.215±0.00
75+ C.A	18.13±0.05	5.46±0.47	0.143±0.00	0.271±0.02	0.201±0.01	0.253±0.00
Control 100	16.66±0.90	6.10±0.33	0.102±0.00	0.134±0.08	0.149±0.05	0.217±0.00
100+ C.A	17.36±0.35	7.30±0.26	0.151±0.01	0.311±0.00	0.200±0.01	0.276±0.00
Control 25	10.17±1.81	4.13±0.55	0.089±0.00	0.125±0.03	0.132±0.00	0.147±0.00
25 + EDTA and C.A	10.26±0.20	5.13±0.05	0.108±0.01	0.130±0.00	0.140±0.00	0.155±0.00
Control 50	12.13±0.96	4.44±0.33	0.092±0.00	0.134±0.08	0.136±0.00	0.151±0.01
50+ EDTA and C.A	13.00±0.72	4.50±0.47	0.110±0.00	0.135±0.00	0.135±0.00	0.160±0.00
Control 75	15.56±2.67	5.21±0.33	0.098±0.00	0.128±0.03	0.143±0.00	0.215±0.00
75+ EDTA and C.A	14.00±0.05	5.80±0.47	0.115±0.00	0.140±0.02	0.140±0.01	0.218±0.00
Control 100	16.66±0.90	6.10±0.33	0.102±0.00	0.134±0.08	0.149±0.05	0.217±0.00
100+ EDTA and C.A	14.36±0.35	7.00±0.26	0.120±0.01	0.145±0.00	0.155±0.01	0.225±0.00

Concentration (20 mM)						
Treatment	**RL(cm)**	**SL(cm)**	**RDW(gm)**	**SDW(gm)**	**RFW (gm)**	**SFW (gm)**
Control	10.03±0.20	7.33±0.05	0.109±0.00	0.232±0.03	0.126±0.00	0.202±0.00
Control 25	12.36±0.25	5.33±0.25	0.084±0.00	0.098±0.04	0.145±0.00	0.162±0.00
25+ EDTA	18.30±0.34	8.70±0.55	0.134±0.00	0.295±0.05	0.213±0.00	0.325±0.00
Control 50	14.26±0.20	8.20±0.20	0.067±0.00	0.143±0.04	0.165±0.00	0.167±0.08
50 + EDTA	19.26±0.15	10.20±0.17	0.136±0.00	0.250±0.00	0.225±0.00	0.336±0.00
Control 75	15.80±0.70	9.23±0.88	0.083±0.00	0.134±0.03	0.155±0.00	0.188±0.09
75+ EDTA	19.30±0.30	11.83±0.37	0.142±0.00	0.270±0.00	0.246±0.00	0.342±0.00
Control 100	23.90±0.65	10.86±0.68	0.085±0.00	0.123±0.04	0.171±0.00	0.178±0.00
100+ EDTA	37.23±0.23	12.16±0.30	0.154±0.00	0.312±0.00	0.259±0.00	0.355±0.00
Control 25	13.26±0.00	5.23±0.88	0.083±0.00	0.132±0.03	0.143±0.00	0.158±0.00
25 + C.A	25.33±0.20	6.20±0.20	0.133±0.01	0.210±0.00	0.190±0.00	0.225±0.00

Control 50	15.50±0.51	5.86±0.68	0.093±0.07	0.135±0.03	0.146±0.00	0.166±0.09
50+ C.A	23.20±0.08	6.33±0.40	0.163±0.00	0.257±0.01	0.204±0.00	0.245±0.00
Control 75	17.86±0.90	6.21±0.00	0.097±0.08	0.138±0.04	0.154±0.00	0.232±0.09
75 + C.A	23.73±4.66	7.56±0.35	0.155±0.02	0.311±0.00	0.212±0.00	0.265±0.00
Control 100	18.33±0.04	7.42±0.67	0.106±0.09	0.140±0.08	0.169±0.00	0.235±0.09
100 + C.A	20.30±0.24	8.43±0.35	0.173±0.00	0.323±0.01	0.216±0.00	0.287±0.00
Control 25	13.26±0.00	5.23±0.88	0.083±0.00	0.132±0.03	0.143±0.00	0.158±0.00
25+ EDTA and C.A	15.00±0.20	6.00±0.20	0.100±0.01	0.140±0.00	0.150±0.00	0.170±0.00
Control 50	15.50±0.51	5.86±0.68	0.093±0.07	0.135±0.03	0.146±0.00	0.166±0.09
50+ EDTA and C.A	16.00±0.08	6.90±0.40	0.115±0.00	0.150±0.01	0.165±0.00	0.185±0.00
Control 75	17.86±0.90	6.21±0.00	0.097±0.08	0.138±0.04	0.154±0.00	0.232±0.09
75+ EDTA and C.A	18.70±4.66	7.00±0.35	0.120±0.02	0.155±0.00	0.180±0.00	0.195±0.00
Control 100	18.33±0.04	7.42±0.67	0.106±0.09	0.140±0.08	0.169±0.00	0.235±0.09
100+ EDTA and C.A	19.00±0.24	8.00±0.35	0.125±0.00	0.165±0.01	0.195±0.00	0.200±0.00

*Individual readings n =3 for each plant were averaged and are presented with ±SD.

*Significant correlation was found between root length (RL), shoot length (SL), root, shoot dry weight (RDW, SDW) and root, shoot fresh weight (RFW, SFW) at 0.05 level of significance;

EDTA: Ethylene diamine tetraacetic acid; C.A: Citric acid.

Table 4.15.2: Biomass of *P. juliflora* plants grown in soil for 60 days in different F concentration viz. control, 25, 50, 75 and 100 mg kg^{-1} NaF soil after treatment with chelates EDTA and CA

Concentration (5 mM)						
Treatment	**RL(cm)**	**SL(cm)**	**RDW(gm)**	**SDW(gm)**	**RFW (gm)**	**SFW (gm)**
Control	15.70±0.51	8.06±0.46	0.136±0.08	0.184±0.01	0.180±0.00	0.252±0.00
Control 25	6.76±0.63	7.45±0.92	0.123±0.01	0.129±0.00	0.137±0.00	0.158±0.00
25+ EDTA	10.33±1.78	7.66±0.28	0.277±0.00	0.281±0.00	0.197±0.00	0.316±0.00
Control 50	7.43±0.05	8.23±0.80	0.155±0.00	0.140±0.00	0.147±0.00	0.170±0.00
50+ EDTA	12.93±3.34	8.36±0.75	0.284±0.00	0.290±0.00	0.203±0.00	0.329±0.01
Control 75	10.83±1.09	8.33±0.80	0.182±0.00	0.143±0.01	0.158±0.00	0.185±0.00
75+ EDTA	13.53±1.78	9.06±0.93	0.285±0.00	0.295±0.00	0.215±0.00	0.345±0.00

Control 100	12.33±0.23	9.80±1.38	0.186±0.00	0.145±0.00	0.181±0.00	0.209±0.01
100+ EDTA	17.03±0.46	10.06±0.92	0.302±0.00	0.299±0.00	0.226±0.00	0.361±0.00
Control 25	8.13±2.82	7.70±2.07	0.121±0.01	0.118±0.00	0.125±0.00	0.153±0.01
25 + C.A	10.80±2.51	13.73±1.32	0.177±0.01	0.188±0.00	0.191±0.00	0.202±0.00
Control 50	12.73±2.65	13.00±0.34	0.129±0.00	0.120±0.00	0.147±0.00	0.170±0.00
50+ C.A	15.73±1.67	14.93±0.23	0.184±0.00	0.226±0.01	0.201±0.00	0.243±0.00
Control 75	13.78±0.90	14.06±1.09	0.143±0.00	0.173±0.00	0.180±0.00	0.198±0.00
75 + C.A	16.36±2.71	17.93±1.50	0.195±0.01	0.238±0.00	0.226±0.00	0.281±0.00
Control 100	14.70±0.51	13.23±0.98	0.173±0.08	0.185±0.00	0.196±0.00	0.239±0.00
100+ C.A	17.36±4.44	19.10±0.51	0.199±0.01	0.244±0.00	0.235±0.00	0.291±0.00
Control 25	8.13±2.82	7.70±2.07	0.121±0.01	0.118±0.00	0.125±0.00	0.153±0.01
25 + EDTA and C.A	9.80±0.00	9.73±1.32	0.150±0.01	0.120±0.00	0.135±0.00	0.160±0.00
Control 50	12.73±2.65	13.00±0.34	0.129±0.00	0.120±0.00	0.147±0.00	0.170±0.00
50 + EDTA and C.A	13.73±1.00	13.93±0.23	0.155±0.00	0.190±0.01	0.150±0.00	0.185±0.00
Control 75	13.78±0.90	14.06±1.09	0.143±0.00	0.173±0.00	0.180±0.00	0.198±0.00
75 + EDTA and C.A	14.36±0.00	15.00±1.50	0.158±0.01	0.195±0.00	0.190±0.00	0.214±0.00
Control 100	14.70±0.51	15.23±0.98	0.173±0.08	0.185±0.00	0.196±0.00	0.239±0.00
100 + EDTA and C.A	15.36±0.0	15.10±0.51	0.160±0.01	0.211±0.00	0.215±0.00	0.250±0.00

Concentration (10 mM)						
Treatment	**RL(cm)**	**SL(cm)**	**RDW(gm)**	**SDW(gm)**	**RFW (gm)**	**SFW (gm)**
Control	15.70±0.51	8.06±0.46	0.136±0.08	0.184±0.01	0.180±0.00	0.252±0.00
Control 25	15.80±0.86	8.03±0.00	0.128±0.00	0.133±0.01	0.147±0.00	0.173±0.00
25+ EDTA	18.90±3.11	12.80±1.81	0.286±0.00	0.286±0.00	0.216±0.00	0.325±0.00
Control 50	16.66±4.44	10.76±1.44	0.162±0.00	0.150±0.09	0.149±0.00	0.185±0.01
50+ EDTA	19.86±0.80	14.13±0.98	0.256±0.00	0.291±0.00	0.225±0.00	0.343±0.01
Control 75	18.56±3.75	12.83±1.50	0.172±0.00	0.146±0.00	0.170±0.00	0.191±0.00
75+ EDTA	19.96±2.82	15.16±0.92	0.282±0.00	0.277±0.01	0.233±0.01	0.352±0.01
Control 100	21.66±6.17	14.16±0.92	0.188±0.00	0.154±0.01	0.184±0.00	0.209±0.01
100 + EDTA	22.23±0.63	16.06±0.75	0.311±0.01	0.292±0.05	0.247±0.00	0.363±0.01
Control 25	7.36±0.32	11.46±0.05	0.132±0.01	0.133±0.00	0.137±0.00	0.159±0.00
25 + C.A	12.46±2.54	14.60±0.11	0.195±0.01	0.201±0.00	0.196±0.00	0.212±0.00
Control 50	12.63±0.05	12.06±0.75	0.135±0.00	0.184±0.00	0.158±0.00	0.181±0.00
50+ C.A	13.80±2.59	15.16±0.92	0.186±0.00	0.232±0.00	0.212±0.00	0.258±0.00
Control 75	13.90±0.80	13.56±2.13	0.150±0.00	0.185±0.00	0.191±0.00	0.215±0.00
75 + C.A	15.13±2.13	16.50±0.69	0.197±0.01	0.240±0.00	0.237±0.00	0.284±0.01

Control 100	14.23±0.07	14.30±2.42	0.168±0.00	0.199±0.01	0.218±0.00	0.239±0.00
100+ C.A	17.66±3.29	16.76±1.09	0.204±0.01	0.250±0.00	0.242±0.00	0.304±0.00
Control 25	7.36±0.32	11.46±0.05	0.132±0.01	0.133±0.00	0.137±0.00	0.159±0.00
25+ EDTA and C.A	9.46±2.54	12.60±0.11	0.145±0.01	0.155±0.00	0.145±0.00	0.170±0.00
Control 50	12.63±0.05	12.06±0.75	0.135±0.00	0.184±0.00	0.158±0.00	0.181±0.00
50 + EDTA and C.A	12.80±2.59	13.16±0.92	0.145±0.00	0.190±0.00	0.160±0.00	0.185±0.00
Control 75	13.90±0.80	13.56±2.13	0.150±0.00	0.185±0.00	0.191±0.00	0.215±0.00
75+ EDTA and C.A	14.13±2.13	14.50±0.69	0.155±0.01	0.195±0.00	0.195±0.00	0.212±0.01
Control 100	14.23±0.07	14.30±2.42	0.168±0.00	0.199±0.01	0.218±0.00	0.239±0.00
100+ EDTA and C.A	11.66±3.29	10.76±1.09	0.170±0.01	0.220±0.00	0.220±0.00	0.250±0.00

Concentration (15 mM)						
Treatment	**RL(cm)**	**SL(cm)**	**RDW(gm)**	**SDW(gm)**	**RFW (gm)**	**SFW (gm)**
Control	15.70±0.51	8.06±0.46	0.136±0.08	0.184±0.01	0.180±0.00	0.252±0.00
Control 25	16.33±0.40	7.53±2.36	0.131±0.05	0.144±0.01	0.158±0.00	0.181±0.00
25 + EDTA	11.73±6.29	8.73±2.02	0.287±0.00	0.295±0.00	0.248±0.00	0.336±0.00
Control 50	18.36±0.28	8.60±1.90	0.170±0.00	0.159±0.00	0.170±0.00	0.191±0.00
50 + EDTA	23.93±1.21	9.56±0.40	0.289±0.00	0.300±0.01	0.253±0.00	0.350±0.01
Control 75	19.13±4.56	9.50±2.42	0.183±0.00	0.154±0.01	0.181±0.00	0.201±0.00
75+ EDTA	24.30±2.77	10.56±0.40	0.296±0.00	0.310±0.00	0.257±0.08	0.359±0.00
Control 100	21.33±2.13	10.30±2.42	0.192±0.00	0.159±0.00	0.191±0.00	0.210±0.01
100+ EDTA	25.70±0.17	12.17±0.86	0.299±0.05	0.318±0.00	0.258±0.00	0.373±0.01
Control 25	15.90±1.94	15.83±1.15	0.137±0.00	0.144±0.00	0.147±0.00	0.162±0.01
25 + C.A	12.86±2.36	16.40±2.07	0.188±0.01	0.216±0.00	0.205±0.01	0.224±0.00
Control 50	16.96±1.23	16.73±2.02	0.143±0.00	0.191±0.00	0.170±0.00	0.192±0.00
50+ C.A	22.43±1.61	19.96±1.61	0.174±0.01	0.242±0.00	0.225±0.00	0.269±0.00
Control 75	17.86±1.61	17.40±2.59	0.159±0.00	0.186±0.00	0.187±0.00	0.233±0.00
75+ C.A	14.43±1.86	20.96±2.65	0.192±0.00	0.246±0.00	0.247±0.00	0.291±0.00
Control 100	19.83±2.13	17.66±4.38	0.178±0.00	0.210±0.01	0.228±0.00	0.248±0.00
100 + C.A	18.40±1.21	23.23±2.19	0.198±0.01	0.266±0.01	0.244±0.00	0.315±0.00
Control 25	15.90±6.94	15.83±1.15	0.137±0.00	0.144±0.00	0.147±0.00	0.162±0.01
25+ EDTA and C.A	16.86±1.36	16.00±2.07	0.140±0.01	0.150±0.00	0.155±0.01	0.170±0.00
Control 50	16.96±1.23	16.73±2.02	0.143±0.00	0.191±0.00	0.170±0.00	0.192±0.00
50+ EDTA and C.A	22.00±1.61	16.75±1.61	0.155±0.01	0.190±0.00	0.180±0.00	0.210±0.00
Control 75	17.86±4.61	17.40±2.59	0.159±0.00	0.186±0.00	0.187±0.00	0.233±0.00

75+ EDTA and C.A	14.00±1.86	18.00±2.65	0.160±0.00	0.196±0.00	0.190±0.00	0.245±0.00
Control 100	19.83±2.13	17.66±1.38	0.178±0.00	0.210±0.01	0.228±0.00	0.248±0.00
100+ EDTA and C.A	17.00±1.21	19.00±2.19	0.180±0.01	0.215±0.01	0.230±0.00	0.250±0.00

Concentration (20 mM)						
Treatment	**RL(cm)**	**SL(cm)**	**RDW(gm)**	**SDW(gm)**	**RFW (gm)**	**SFW (gm)**
Control	15.70±0.51	8.06±0.46	0.136±0.08	0.184±0.01	0.180±0.00	0.252±0.00
Control 25	17.77±0.17	7.50±1.90	0.132±0.00	0.150±0.01	0.176±0.00	0.192±0.00
25 + EDTA	19.83±2.30	11.10±2.17	0.267±0.01	0.314±0.00	0.260±0.00	0.348±0.00
Control 50	19.60±0.69	8.66±0.40	0.181±0.00	0.180±0.00	0.187±0.00	0.204±0.01
50+ EDTA	20.60±2.51	11.72±1.38	0.277±0.01	0.317±0.00	0.261±0.00	0.362±0.01
Control 75	21.56±1.15	10.13±3.92	0.188±0.00	0.182±0.00	0.196±0.01	0.219±0.01
75+ EDTA	21.46±3.75	12.80±0.69	0.294±0.00	0.322±0.01	0.263±0.00	0.363±0.01
Control 100	24.23±1.44	7.73±2.02	0.216±0.00	0.185±0.00	0.198±0.00	0.229±0.01
100 + EDTA	36.86±1.09	11.83±2.36	0.320±0.00	0.327±0.00	0.281±0.00	0.380±0.00
Control 25	17.50±5.19	18.16±1.44	0.147±0.00	0.154±0.00	0.181±0.00	0.180±0.00
25+ C.A	21.26±1.32	21.80±1.21	0.196±0.01	0.232±0.00	0.221±0.00	0.238±0.00
Control 50	18.36±3.92	19.00±2.83	0.147±0.00	0.199±0.01	0.201±0.01	0.196±0.00
50+ C.A	22.22±0.00	22.86±1.78	0.203±0.01	0.247±0.00	0.237±0.00	0.282±0.00
Control 75	19.00±2.25	21.60±5.96	0.173±0.00	0.208±0.02	0.239±0.00	0.245±0.00
75+ C.A	22.83±1.44	24.96±3.17	0.215±0.02	0.255±0.00	0.258±0.00	0.303±0.00
Control 100	22.83±3.00	22.60±5.02	0.188±0.01	0.215±0.01	0.235±0.00	0.259±0.00
100+ C.A	24.00±2.07	25.83±1.15	0.225±0.01	0.287±0.00	0.247±0.00	0.329±0.01
Control 25	17.50±2.19	18.16±1.44	0.147±0.00	0.154±0.00	0.181±0.00	0.180±0.00
25+ EDTA and C.A	19.26±1.32	19.40±1.21	0.156±0.01	0.165±0.00	0.195±0.00	0.190±0.00
Control 50	18.36±3.92	19.00±2.83	0.147±0.00	0.199±0.01	0.201±0.01	0.196±0.00
50 + DTA and C.A	19.00±0.00	18.00±0.00	0.175±0.00	0.220±0.00	0.215±0.00	0.212±0.00
Control 75	19.00±2.25	21.60±2.96	0.173±0.00	0.208±0.02	0.239±0.00	0.245±0.00
75+ EDTA and C.A	20.00±0.00	21.96±0.00	0.185±0.00	0.225±0.00	0.220±0.00	0.250±0.00
Control 100	22.83±3.00	22.60±5.02	0.188±0.01	0.215±0.01	0.235±0.00	0.259±0.00
100+ EDTA and C.A	22.00±0.00	23.83±0.00	0.195±0.00	0.235±0.00	0.235±0.00	0.270±0.00

*Individual readings n =3 for each plant were averaged and are presented with ±SD.

*Significant correlation was found between root length (RL), shoot length (SL), root, shoot dry weight (RDW, SDW) and root, shoot fresh weight (RFW, SFW) at 0.05 level of significance;

EDTA: Ethylene diamine tetraacetic acid; C.A: Citric acid.

Table 4.15.3: Biomass of *P. juliflora* plants grown in soil for 90 days in different F concentration viz. control, 25, 50, 75 and 100 mg kg^{-1} NaF soil after treatment with chelates EDTA and CA

Concentration (5 mM)						
Treatment	**RL(cm)**	**SL(cm)**	**RDW(gm)**	**SDW(gm)**	**RFW (gm)**	**SFW (gm)**
Control	16.40±0.23	10.03±3.86	0.165±0.01	0.191±0.00	0.226±0.00	0.259±0.00
Control 25	18.60±0.17	12.80±0.34	0.143±0.00	0.143±0.00	0.178±0.00	0.218±0.00
25+ EDTA	20.10±1.21	15.26±2.13	0.286±0.00	0.282±0.02	0.226±0.00	0.327±0.00
Control 50	19.90±0.69	14.76±1.84	0.176±0.00	0.155±0.00	0.181±0.00	0.192±0.00
50+ EDTA	22.43±0.09	16.10±1.57	0.287±0.00	0.313±0.00	0.235±0.00	0.337±0.02
Control 75	22.93±1.44	15.80±1.73	0.203±0.00	0.176±0.03	0.191±0.00	0.197±0.00
75+ EDTA	25.03±0.75	16.33±5.57	0.315±0.00	0.279±0.01	0.246±0.00	0.376±0.00
Control 100	23.20±1.03	16.33±1.27	0.211±0.00	0.188±0.00	0.204±0.00	0.180±0.00
100+ EDTA	28.13±1.09	18.90±2.94	0.324±0.00	0.277±0.00	0.259±0.00	0.392±0.00
Control 25	16.06±1.15	12.23±0.55	0.134±0.00	0.138±0.00	0.159±0.00	0.191±0.00
25 + C.A	17.06±1.54	15.96±2.54	0.184±0.00	0.197±0.00	0.227±0.00	0.236±0.00
Control 50	17.46±1.15	14.66±0.80	0.141±0.00	0.177±0.00	0.169±0.00	0.232±0.03
50 + C.A	18.03±0.75	16.10±2.25	0.190±0.00	0.223±0.01	0.248±0.00	0.254±0.00
Control 75	18.23±1.09	15.40±1.21	0.166±0.01	0.188±0.00	0.187±0.00	0.237±0.09
75+ C.A	19.60±0.90	18.16±0.63	0.225±0.01	0.248±0.00	0.263±0.01	0.291±0.00
Control 100	22.66±0.05	16.10±0.51	0.187±0.00	0.214±0.00	0.196±0.00	0.284±0.00
100 + C.A	23.90±0.69	18.76±0.63	0.236±0.00	0.255±0.00	0.270±0.00	0.293±0.00
Control 25	16.06±1.15	12.23±0.55	0.134±0.00	0.138±0.00	0.159±0.00	0.191±0.00
25+ EDTA and C.A	16.06±1.54	12.96±0.00	0.144±0.00	0.147±0.00	0.165±0.00	0.200±0.00
Control 50	17.46±1.15	14.66±0.80	0.141±0.00	0.177±0.00	0.169±0.00	0.232±0.03
50+ EDTA and C.A	17.03±0.75	15.10±0.00	0.150±0.00	0.155±0.01	0.170±0.00	0.235±0.00
Control 75	18.23±1.09	15.40±1.21	0.166±0.01	0.188±0.00	0.187±0.00	0.237±0.09
75 + EDTA and C.A	18.60±0.90	18.16±0.00	0.175±0.01	0.195±0.00	0.185±0.01	0.254±0.00
Control 100	22.66±0.05	16.10±0.51	0.187±0.00	0.214±0.00	0.196±0.00	0.284±0.00
10 + EDTA and C.A	20.00±0.69	17.00±0.00	0.198±0.00	0.220±0.00	0.210±0.00	0.290±0.00

Concentration (10 mM)						
Treatment	**RL(cm)**	**SL(cm)**	**RDW(gm)**	**SDW(gm)**	**RFW (gm)**	**SFW (gm)**
Control	16.40±0.23	10.03±3.86	0.165±0.01	0.191±0.00	0.226±0.00	0.259±0.00
Control 25	19.96±0.63	13.10±0.69	0.153±0.00	0.149±0.00	0.189±0.00	0.218±0.00
25+ EDTA	23.86±2.36	15.30±2.51	0.301±0.00	0.285±0.00	0.234±0.00	0.332±0.00
Control 50	20.36±1.03	14.23±2.82	0.177±0.00	0.162±0.01	0.188±0.01	0.187±0.00
50+ EDTA	24.46±1.15	15.60±1.27	0.298±0.01	0.315±0.00	0.248±0.00	0.348±0.00
Control 75	22.16±0.46	15.43±1.27	0.208±0.00	0.181±0.00	0.196±0.01	0.206±0.01
75 + EDTA	25.76±1.84	16.56±1.87	0.316±0.00	0.288±0.00	0.258±0.00	0.381±0.00
Control 100	24.43±1.09	16.70±1.84	0.228±0.00	0.189±0.00	0.218±0.00	0.237±0.00
100+ EDTA	28.23±1.27	16.90±0.11	0.328±0.00	0.291±0.00	0.274±0.01	0.395±0.01
Control 25	17.43±1.09	14.46±0.05	0.138±0.00	0.134±0.00	0.170±0.00	0.190±0.00
25+ C.A	18.13±1.27	15.86±3.92	0.190±0.00	0.203±0.00	0.238±0.00	0.237±0.00
Control 50	18.60±1.38	15.23±1.27	0.145±0.00	0.184±0.00	0.175±0.01	0.198±0.00
50+ C.A	19.40±2.50	16.06±0.63	0.222±0.00	0.237±0.00	0.259±0.00	0.270±0.00
Control 75	19.11±0.40	17.16±0.63	0.165±0.00	0.191±0.00	0.191±0.00	0.237±0.00
75 + C.A	22.60±1.78	18.10±2.07	0.229±0.00	0.256±0.00	0.273±0.01	0.303±0.00
Control 100	21.60±1.38	19.10±0.69	0.192±0.00	0.216±0.00	0.225±0.00	0.248±0.00
100+ C.A	23.16±1.32	19.86±0.80	0.240±0.00	0.266±0.00	0.274±0.01	0.306±0.01
Control 25	17.43±1.09	14.46±0.05	0.138±0.00	0.134±0.00	0.170±0.00	0.190±0.00
25+ EDTA and C.A	18.00±0.00	12.86±0.00	0.140±0.00	0.140±0.00	0.180±0.00	0.215±0.00
Control 50	18.60±1.38	15.23±1.27	0.145±0.00	0.184±0.00	0.175±0.01	0.198±0.00
50+ EDTA and C.A	19.40±0.00	16.00±0.00	0.152±0.00	0.190±0.00	0.195±0.00	0.220±0.00
Control 75	19.11±0.40	17.16±0.63	0.165±0.00	0.191±0.00	0.191±0.00	0.237±0.00
75+ EDTA and C.A	10.60±0.00	17.10±0.00	0.170±0.00	0.195±0.00	0.210±0.01	0.240±0.00
Control 100	21.60±1.38	19.10±0.69	0.192±0.00	0.216±0.00	0.225±0.00	0.248±0.00
100+ EDTA and C.A	22.16±0.00	20.00±0.00	0.210±0.00	0.220±0.00	0.230±0.01	0.250±0.01

Concentration (15 mM)						
Treatment	**RL(cm)**	**SL(cm)**	**RDW(gm)**	**SDW(gm)**	**RFW (gm)**	**SFW (gm)**
Control	16.40±0.23	10.03±1.86	0.165±0.01	0.191±0.00	0.226±0.00	0.259±0.00
Control 25	18.46±1.15	15.66±1.32	0.155±0.00	0.150±0.00	0.180±0.00	0.220±0.00
25+ EDTA	19.70±1.27	15.53±1.96	0.288±0.00	0.293±0.00	0.249±0.00	0.342±0.00
Control 50	19.00±0.51	16.63±3.34	0.186±0.00	0.172±0.01	0.192±0.00	0.215±0.00
50 + EDTA	20.80±0.86	17.70±1.05	0.297±0.01	0.328±0.01	0.273±0.01	0.359±0.00
Control 75	19.23±0.57	17.43±2.65	0.224±0.10	0.185±0.00	0.232±0.00	0.226±0.00
75+ EDTA	23.80±0.51	18.16±6.92	0.320±0.00	0.396±0.01	0.274±0.01	0.390±0.00

Control 100	20.80±0.00	19.66±1.67	0.234±0.01	0.191±0.00	0.237±0.00	0.255±0.00
100 + EDTA	24.46±2.71	20.10±1.01	0.334±0.00	0.398±0.01	0.285±0.01	0.401±0.01
Control 25	20.36±0.80	15.20±2.25	0.143±0.00	0.148±0.00	0.178±0.00	0.196±0.00
25 + C.A	20.70±1.90	15.96±0.63	0.221±0.00	0.219±0.01	0.248±0.00	0.257±0.00
Control 50	22.03±0.57	15.86±1.67	0.148±0.00	0.185±0.00	0.184±0.01	0.237±0.00
50+ C.A	23.46±2.02	16.13±1.86	0.216±0.00	0.248±0.00	0.270±0.00	0.281±0.00
Control 75	23.23±1.78	16.76±1.50	0.166±0.00	0.199±0.01	0.198±0.00	0.248±0.00
75+ C.A	25.83±1.08	17.83±2.30	0.235±0.00	0.270±0.00	0.276±0.00	0.315±0.00
Control 100	24.00±2.94	17.36±2.02	0.202±0.00	0.227±0.00	0.234±0.00	0.256±0.00
100 + C.A	26.26±0.80	19.93±0.75	0.244±0.00	0.274±0.00	0.281±0.00	0.324±0.00
Control 25	20.36±0.80	15.20±2.25	0.143±0.00	0.148±0.00	0.178±0.00	0.196±0.00
25 + EDTA and C.A	20.00±0.00	15.00±0.00	0.155±0.00	0.160±0.01	0.180±0.00	0.200±0.00
Control 50	22.03±0.57	15.86±1.67	0.148±0.00	0.185±0.00	0.184±0.01	0.237±0.00
50 + EDTA and C.A	23.00±0.00	16.00±0.00	0.160±0.00	0.190±0.00	0.195±0.00	0.235±0.00
Control 75	23.23±1.78	16.76±1.50	0.166±0.00	0.199±0.01	0.198±0.00	0.248±0.00
75 + EDTA and C.A	24.00±0.00	17.00±0.00	0.170±0.00	0.210±0.00	0.220±0.00	0.260±0.00
Control 100	24.00±2.94	17.36±2.02	0.202±0.00	0.227±0.00	0.234±0.00	0.256±0.00
100 + EDTA and C.A	25.00±0.00	18.00±0.00	0.185±0.00	0.230±0.00	0.240±0.00	0.270±0.00

Concentration (20 mM)						
Treatment	**RL(cm)**	**SL(cm)**	**RDW(gm)**	**SDW(gm)**	**RFW (gm)**	**SFW (gm)**
Control	16.40±0.23	10.03±3.86	0.163±0.01	0.191±0.00	0.226+0.00	0.259±0.00
Control 25	23.40±0.86	16.90±2.00	0.159±0.00	0.148±0.00	0.191±0.00	0.220±0.01
25+ EDTA	23.20±1.21	18.80±2.00	0.312±0.00	0.305±0.00	0.263±0.00	0.358±0.00
Control 50	23.20±1.38	17.43±1.38	0.187±0.00	0.183±0.01	0.201±0.00	0.226±0.00
50+ EDTA	24.33±1.10	18.00±1.15	0.315±0.01	0.332±0.00	0.281±0.00	0.376±0.00
Control 75	24.03±0.63	17.86±1.27	0.226±0.00	0.191±0.00	0.242±0.00	0.237±0.00
75+ EDTA	26.10±0.51	19.20±0.69	0.327±0.00	0.339±0.01	0.281±0.00	0.403±0.09
Control 100	28.23±0.57	18.73±3.17	0.237±0.00	0.198±0.01	0.256±0.00	0.276±0.00
100+ EDTA	35.36±1.32	22.20±0.69	0.340±0.00	0.343±0.00	0.296±0.01	0.409±0.03
Control 25	21.10±1.03	17.13±3.23	0.155±0.00	0.159±0.00	0.191±0.00	0.227±0.00
25+ C.A	22.86±1.09	18.76±0.05	0.230±0.00	0.234±0.00	0.259±0.00	0.273±0.00
Control 50	22.23±0.57	18.17±0.00	0.158±0.00	0.196±0.01	0.209±0.01	0.256±0.00
50+ C.A	24.76±0.23	19.23±2.02	0.232±0.00	0.266±0.00	0.281±0.00	0.292±0.00
Control 75	24.33±3.00	19.63±0.05	0.175±0.00	0.203±0.01	0.221±0.00	0.275±0.00
75+ C.A	26.73±1.80	22.10±2.69	0.246±0.00	0.277±0.00	0.288±0.00	0.346±0.00
Control 100	26.73±2.48	20.90±2.42	0.214±0.00	0.243±0.00	0.243±0.01	0.270±0.00

100 + C.A	28.03±0.51	24.76±0.11	0.254±0.00	0.298±0.00	0.292±0.00	0.343±0.00
Control 25	21.10±1.03	17.13±1.23	0.155±0.00	0.159±0.00	0.191±0.00	0.227±0.00
25+ EDTA and C.A	24.86±1.09	17.50±0.05	0.165±0.00	0.170±0.00	0.220±0.00	0.230±0.00
Control 50	22.23±0.57	18.17±0.00	0.158±0.00	0.196±0.01	0.209±0.01	0.256±0.00
50+ EDTA and C.A	25.76±0.23	19.00±2.02	0.165±0.00	0.210±0.00	0.225±0.00	0.255±0.00
Control 75	24.33±3.00	19.63±0.05	0.175±0.00	0.203±0.01	0.221±0.00	0.275±0.00
75+ EDTA and C.A	26.00±1.80	20.00±2.69	0.185±0.00	0.220±0.00	0.230±0.00	0.285±0.00
Control 100	26.73±2.48	20.90±2.42	0.214±0.00	0.243±0.00	0.243±0.01	0.270±0.00
100+ EDTA and C.A	27.00±0.51	21.00±0.11	0.220±0.00	0.250±0.00	0.250±0.00	0.290±0.00

*Individual readings n =3 for each plant were averaged and are presented with ±SD.

*Significant correlation was found between root length (RL), shoot length (SL), root, shoot dry weight (RDW, SDW) and root, shoot fresh weight (RFW, SFW) at 0.05 level of significance;

EDTA: Ethylene diamine tetraacetic acid; C.A: Citric acid.

Table 4.15.4: Biomass of *P. juliflora* plants grown in soil for 120 days in different F concentration viz. control, 25, 50, 75 and 100 mg kg^{-1} NaF soil after treatment with chelates EDTA and CA

Concentration (5 mM)						
Treatment	**RL(cm)**	**SL(cm)**	**RDW(gm)**	**SDW(gm)**	**RFW (gm)**	**SFW (gm)**
Control	24.16±1.27	15.96±1.27	0.212±0.00	0.188±0.00	0.229±0.01	0.266±0.00
Control 25	24.46±0.57	20.93±2.36	0.166±0.00	0.157±0.00	0.190±0.00	0.250±0.01
25+ EDTA	26.30±0.69	22.80±1.03	0.313±0.00	0.311±0.00	0.237±0.00	0.339±0.00
Control 50	23.50±0.69	19.76±3.34	0.166±0.00	0.146±0.00	0.181±0.00	0.148±0.00
50+ EDTA	28.50±0.69	23.96±1.27	0.324±0.00	0.314±0.00	0.249±0.00	0.349±0.00
Control 75	22.06±1.27	18.93±1.44	0.156±0.00	0.137±0.00	0.180±0.02	0.246±0.01
75+ EDTA	32.10±1.03	24.17±0.69	0.333±0.00	0.336±0.06	0.258±0.00	0.376±0.00
Control 100	20.13±0.63	16.73±1.13	0.147±0.00	0.134±0.00	0.175±0.01	0.234±0.01
100+ EDTA	35.33±0.80	27.83±1.84	0.349±0.00	0.356±0.00	0.271±0.00	0.406±0.01
Control 25	19.00±1.38	12.30±2.07	0.174±0.01	0.126±0.00	0.210±0.02	0.187±0.00
25 + C.A	19.26±0.75	17.30±0.69	0.214±0.00	0.233±0.00	0.239±0.00	0.247±0.00
Control 50	14.80±0.86	11.86±1.09	0.170±0.00	0.121±0.00	0.203±0.00	0.178±0.00
50 + C.A	22.56±0.57	18.00±1.21	0.227±0.00	0.255±0.00	0.255±0.00	0.261±0.01

Control 75	12.30±0.69	10.66±3.34	0.170±0.00	0.120±0.00	0.201±0.00	0.174±0.00
75+ C.A	23.20±0.34	19.20±0.46	0.251±0.01	0.270±0.00	0.281±0.00	0.308±0.01
Control 100	11.33±0.63	19.73±0.05	0.166±0.00	0.120±0.00	0.187±0.00	0.163±0.01
100 + C.A	24.36±0.57	22.30±2.69	0.248±0.00	0.285±0.00	0.282±0.00	0.307±0.01
Control 25	19.00±1.38	12.30±2.07	0.174±0.01	0.126±0.00	0.210±0.02	0.187±0.00
25+ EDTA and C.A	19.10±0.75	15.00±0.69	0.175±0.00	0.140±0.00	0.220±0.00	0.190±0.00
Control 50	14.80±0.86	1.86±1.09	0.170±0.00	0.121±0.00	0.203±0.00	0.178±0.00
50+ EDTA and C.A	19.00±0.57	16.00±1.21	0.180±0.00	0.150±0.00	0.225±0.00	0.220±0.01
Control 75	12.30±0.69	10.66±3.34	0.170±0.00	0.120±0.00	0.201±0.00	0.174±0.00
75 + EDTA and C.A	22.00±0.34	18.20±0.46	0.190±0.01	0.220±0.00	0.235±0.00	0.260±0.01
Control 100	11.33±0.63	09.73±0.05	0.166±0.00	0.110±0.00	0.187±0.00	0.163±0.01
100+ EDTA and C.A	24.00±0.57	21.00±2.69	0.190±0.00	0.245±0.00	0.230±0.00	0.270±0.01

Concentration (10 mM)						
Treatment	**RL(cm)**	**SL(cm)**	**RDW(gm)**	**SDW(gm)**	**RFW (gm)**	**SFW (gm)**
Control	24.16±1.27	15.96±1.27	0.212±0.00	0.188±0.00	0.229±0.01	0.266±0.00
Control 25	21.50±0.69	21.23±2.02	0.173±0.00	0.156±0.00	0.194±0.00	0.255±0.01
25 + EDTA	27.46±0.75	21.46±0.57	0.325±0.00	0.322±0.00	0.241±0.00	0.344±0.01
Control 50	23.70±1.90	20.30±2.42	0.172±0.00	0.157±0.00	0.191±0.00	0.255±0.00
50+ EDTA	30.60±1.55	22.16±1.27	0.344±0.00	0.325±0.02	0.264±0.00	0.356±0.00
Control 75	22.50±0.69	19.83±2.13	0.167±0.01	0.156±0.00	0.184±0.07	0.249±0.01
75+ EDTA	32.56±1.44	24.33±1.50	0.343±0.00	0.348±0.00	0.268±0.00	0.387±0.00
Control 100	21.13±0.63	18.56±5.77	0.160±0.00	0.155±0.00	0.173±0.01	0.242±0.01
100+ EDTA	37.83±1.84	25.96±1.96	0.358±0.00	0.358±0.00	0.286±0.00	0.419±0.01
Control 25	20.16±0.40	16.53±1.84	0.161±0.01	0.165±0.02	0.181±0.00	0.203±0.00
25+ C.A	21.73±0.11	19.00±1.21	0.217±0.00	0.237±0.00	0.253±0.01	0.252±0.01
Control 50	18.00±0.69	14.43±0.63	0.160±0.00	0.163±0.00	0.180±0.00	0.200±0.02
50+ C.A	25.26±0.57	22.50±0.69	0.237±0.00	0.266±0.00	0.273±0.01	0.281±0.05
Control 75	14.80±0.17	13.30±2.07	0.155±0.00	0.146±0.00	0.179±0.00	0.195±0.01
75+ C.A	26.83±0.11	24.90±5.19	0.253±0.02	0.281±0.00	0.290±0.00	0.321±0.01
Control 100	12.40±1.90	12.00±0.36	0.150±0.00	0.139±0.00	0.165±0.00	0.187±0.00
100+ C.A	28.56±0.11	25.40±0.86	0.258±0.00	0.288±0.00	0.289±0.00	0.329±0.01
Control 25	20.16±0.40	16.53±1.84	0.161±0.01	0.165±0.02	0.181±0.00	0.203±0.00
25+ EDTA and C.A	17.73±0.11	15.00±1.21	0.170±0.00	0.170±0.00	0.195±0.01	0.210±0.01

Control 50	18.00±0.69	14.43±0.63	0.160±0.00	0.163±0.00	0.180±0.00	0.200±0.02
50+ EDTA and C.A	21.26±0.57	18.50±0.69	0.185±0.00	0.185±0.00	0.211±0.01	0.220±0.05
Control 75	14.80±0.17	13.30±2.07	0.155±0.00	0.146±0.00	0.179±0.00	0.195±0.01
75+ EDTA and C.A	22.83±0.11	21.90±2.19	0.190±0.02	0.190±0.00	0.220±0.00	0.230±0.01
Control 100	12.40±1.90	12.00±0.36	0.150±0.00	0.139±0.00	0.165±0.00	0.187±0.00
100+ EDTA and C.A	22.56±0.11	22.40±0.86	0.200±0.00	0.195±0.00	0.225±0.00	0.300±0.01

Concentration (15 mM)						
Treatment	**RL(cm)**	**SL(cm)**	**RDW(gm)**	**SDW(gm)**	**RFW (gm)**	**SFW (gm)**
Control	24.16±1.27	15.96±1.27	0.212±0.00	0.188±0.00	0.229±0.01	0.266±0.00
Control 25	25.73±1.67	22.40±1.90	0.181±0.00	0.166±0.00	0.210±0.00	0.265±0.01
25+EDTA	28.00±0.69	24.36±0.75	0.342±0.00	0.336±0.00	0.258±0.00	0.351±0.01
Control 50	25.66±1.78	21.43±2.54	0.180±0.00	0.162±0.00	0.206±0.01	0.261±0.00
50+ EDTA	30.06±1.09	28.13±3.05	0.356±0.00	0.334±0.00	0.287±0.00	0.367±0.00
Control 75	24.50±1.38	18.76±3.69	0.178±0.00	0.161±0.00	0.202±0.00	0.260±0.01
75+ EDTA	35.33±0.80	28.46±1.44	0.369±0.00	0.355±0.00	0.290±0.00	0.402±0.00
Control 100	23.43±0.63	16.13±1.79	0.175±0.00	0.160±0.01	0.201±0.01	0.255±0.02
100+EDTA	38.73±0.11	29.00±2.42	0.368±0.00	0.369±0.00	0.306±0.01	0.422±0.00
Control 25	22.16±0.75	19.13±1.32	0.169±0.00	0.181±0.00	0.188±0.00	0.204±0.00
25+ C.A	25.23±0.28	22.46±0.75	0.225±0.00	0.242±0.00	0.259±0.00	0.272±0.01
Control 50	20.03±0.57	18.86±0.92	0.165±0.00	0.180±0.00	0.183±0.00	0.200±0.00
50+C.A	27.60±0.17	22.50±2.25	0.247±0.00	0.277±0.00	0.281±0.00	0.288±0.00
Control 75	17.50±0.17	16.36±2.48	0.163±0.00	0.176±0.01	0.174±0.00	0.193±0.01
75+C.A	28.43±0.63	24.63±5.31	0.256±0.01	0.290±0.00	0.294±0.00	0.329±0.01
Control 100	16.66±1.09	15.96±0.63	0.154±0.00	0.175±0.00	0.172±0.00	0.187±0.01
100+ C.A	32.66±1.27	28.10±1.38	0.266±0.00	0.302±0.00	0.302±0.02	0.341±0.02
Control 25	22.16±0.75	19.13±1.32	0.169±0.00	0.181±0.00	0.188±0.00	0.204±0.00
25+ EDTA and C.A	23.23±0.00	20.00±0.00	0.175±0.00	0.192±0.00	0.190±0.00	0.212±0.00
Control 50	20.03±0.57	18.86±0.92	0.165±0.00	0.180±0.00	0.183±0.00	0.200±0.00
50+ EDTA and C.A	24.60±0.00	21.00±0.00	0.195±0.00	0.200±0.00	0.220±0.00	0.250±0.00
Control 75	17.50±0.17	16.36±2.48	0.163±0.00	0.176±0.01	0.174±0.00	0.193±0.01
75+ EDTA and C.A	28.43±0.00	23.00±0.00	0.210±0.01	0.260±0.00	0.235±0.00	0.270±0.00
Control 100	16.66±1.09	15.96±0.63	0.154±0.00	0.175±0.00	0.172±0.00	0.187±0.01

100+ EDTA and C.A	30.66±0.00	25.00±0.00	0.220±0.00	0.270±0.00	0.250±0.00	0.280±0.00

Concentration (20 mM)						
Treatment	**RL(cm)**	**SL(cm)**	**RDW(gm)**	**SDW(gm)**	**RFW (gm)**	**SFW (gm)**
Control	24.16±1.27	15.96±1.27	0.212±0.00	0.188±0.00	0.229±0.01	0.266±0.00
Control 25	26.00±1.38	23.36±1.78	0.197±0.00	0.173±0.00	0.216±0.01	0.275±0.00
25+ EDTA	30.40±0.69	25.20±2.07	0.358±0.00	0.348±0.00	0.269±0.00	0.372±0.01
Control 50	25.06±4.56	20.76±1.96	0.192±0.00	0.170±0.02	0.210±0.01	0.265±0.00
50+ EDTA	32.00±1.33	26.00±0.69	0.362±0.00	0.345±0.00	0.295±0.01	0.381±0.00
Control 75	24.43±1.84	20.73±0.80	0.182±0.00	0.165±0.00	0.209±0.00	0.260±0.01
75+ EDTA	35.50±0.69	26.26±0.57	0.381±0.00	0.369±0.00	0.296±0.01	0.417±0.00
Control 100	24.50±1.38	19.53±5.31	0.179±0.00	0.160±0.00	0.203±0.01	0.260±0.01
100 + EDTA	44.56±0.57	30.26±0.57	0.385±0.00	0.381±0.00	0.317±0.00	0.428±0.00
Control 25	25.50±1.03	22.66±1.96	0.181±0.00	0.190±0.00	0.207±0.01	0.229±0.00
25+ C.A	35.16±0.57	24.60±1.55	0.237±0.00	0.253±0.00	0.273±0.01	0.288±0.02
Control 50	24.00±0.17	19.03±0.80	0.180±0.00	0.190±0.00	0.205±0.00	0.226±0.00
50+ C.A	36.30±0.69	23.90±1.21	0.256±0.00	0.288±0.00	0.299±0.01	0.313±0.02
Control 75	23.50±0.63	18.50±2.42	0.179±0.00	0.186±0.00	0.204±0.00	0.225±0.01
75 + C.A	38.26±0.40	24.23±0.80	0.266±0.01	0.300±0.00	0.310±0.02	0.362±0.01
Control 100	21.60±0.86	16.50±0.69	0.176±0.00	0.174±0.00	0.201±0.00	0.219±0.00
100 + C.A	38.70±5.36	27.50±0.69	0.277±0.00	0.314±0.00	0.313±0.02	0.357±0.01
Control 25	25.50±1.03	22.66±1.96	0.181±0.00	0.190±0.00	0.207±0.01	0.229±0.00
25 + EDTA and C.A	27.16±0.00	23.00±0.00	0.215±0.00	0.220±0.00	0.220±0.01	0.240±0.02
Control 50	24.00±0.17	19.03±0.80	0.180±0.00	0.190±0.00	0.205±0.00	0.226±0.00
50 + EDTA and C.A	28.30±0.00	24.00±1.21	0.220±0.00	0.235±0.00	0.240±0.01	0.270±0.02
Control 75	23.50±0.63	18.50±2.42	0.179±0.00	0.186±0.00	0.204±0.00	0.225±0.01
75 + EDTA and C.A	30.26±0.00	25.00±0.80	0.225±0.01	0.270±0.00	0.250±0.02	0.280±0.01
Control 100	21.60±0.86	16.50±0.69	0.172±0.00	0.170±0.00	0.201±0.00	0.209±0.00
100 + EDTA and C.A	32.70±0.00	26.00±0.69	0.235±0.00	0.280±0.00	0.260±0.02	0.285±0.01

*Individual readings n =3 for each plant were averaged and are presented with ±SD.

*Significant correlation was found between root length (RL), shoot length

(SL), root, shoot dry weight (RDW, SDW) and root, shoot fresh weight (RFW, SFW) at 0.05 level of significance;

EDTA: Ethylene diamine tetraacetic acid; C.A: Citric acid.

Table 4.16.1: Chlorophyll content of *P. juliflora* seedlings grown in soil for 30 days in different concentrations of F viz. control, 25, 50, 75 and 100 mg kg^{-1} NaF soil after treatment with chelates EDTA and CA

Chlorophyll content (mg g^{-1}fw)								
5 mMKg^{-1}					**10 mMKg^{-1}**			
Treatment	**Chlorophyll a**	**Chlorophyll b**	**Total chlorophyll**	**Chlorophyll a/b**	**Chlorophyll a**	**Chlorophyll b**	**Total chlorophyll**	**Chlorophyll a/b**
Control	4.090±0.13	3.553±0.00	7.643±0.11	1.151	4.090±0.00	3.553±0.34	7.643±0.00	1.151
Control 25	1.009±0.10	0.420±0.00	1.429±0.10	2.402	1.009±0.11	0.420±0.44	1.429±0.21	2.402
25+ EDTA	1.049±0.00	0.621±0.00	1.670±0.21	1.688	1.148±0.19	0.599±0.21	1.747±0.10	1.916
Control 50	0.886±0.00	0.199±0.08	1.085±0.10	4.452	0.886±0.10	0.199±0.87	1.085±0.22	4.452
50+ EDTA	1.130±0.01	0.750±0.18	1.880±0.00	1.506	1.137±0.00	0.755±0.00	1.892±0.18	1.505
Control 75	0.835±0.07	0.382±0.00	1.217±0.19	2.185	0.835±0.11	0.382±0.38	1.217±0.00	2.184
75+ EDTA	1.129±0.00	0.796±0.06	1.925±0.16	1.418	1.146±0.04	0.762±0.11	1.908±0.17	1.503
Control 100	0.729±0.00	0.223±0.07	0.952±0.18	3.269	0.729±0.10	0.223±0.00	0.952±0.10	3.260
100 + EDTA	1.084±0.00	0.720±0.30	1.804±0.00	1.505	1.094±0.10	0.708±0.19	1.802±0.00	1.545
Control 25	0.713±0.12	0.402±0.10	1.115±0.15	1.773	0.713±0.18	0.402±0.15	1.115±0.00	1.773
25 + C.A	1.062±0.18	0.557±0.00	1.619±0.10	1.906	1.068±0.00	0.552±0.18	1.620±0.19	1.935
Control 50	0.849±0.11	0.263±0.19	1.112±0.00	3.228	0.849±0.10	0.263±0.16	1.112±0.14	3.220
50 + C.A	1.014±0.04	0.552±0.71	1.566±0.16	1.836	1.018±0.34	0.553±0.15	1.572±0.18	1.840
Control 75	0.672±0.01	0.214±0.00	0.886±0.10	3.140	0.672±0.07	0.214±0.13	0.886±0.00	3.131
75 + C.A	0.979±0.00	0.539±0.18	1.518±0.11	1.816	0.989±0.00	0.534±0.16	1.523±0.11	1.850
Control 100	0.656±0.00	0.171±0.05	0.827±0.00	3.836	0.656±0.10	0.171±0.18	0.827±0.00	3.836
100 + C.A	1.021±0.10	0.516±0.16	1.537±0.18	1.990	1.020±0.15	0.504±0.00	1.524±0.00	2.021
Control 25	0.713±0.18	0.402±0.18	1.115±0.18	1.773	0.713±0.16	0.402±0.27	1.115±0.22	1.773

25 + EDTA and C.A	1.027±0.19	0.513±0.00	1.540±0.19	2.001	0.593±0.16	0.327±0.17	0.920±0.04	1.813
Control 50	0.849±0.00	0.263±0.17	1.112±0.10	3.228	0.849±0.16	0.263±0.13	1.112±0.05	3.228
50 + EDTA and C.A	0.979±0.17	0.539±0.08	1.518±0.10	1.816	0.989±0.11	0.534±0.16	1.523±0.00	1.852
Control 75	0.672±0.15	0.214±0.08	0.886±0.00	3.140	0.672±0.19	0.214±0.16	0.886±0.19	3.131
75 + EDTA and C.A	1.129±0.10	0.796±0.15	1.925±0.17	1.418	1.140±0.16	0.762±0.16	1.902±0.00	1.500
Control 100	0.656±0.00	0.171±0.01	0.827±0.18	3.836	0.656±0.00	0.171±0.10	0.827±0.18	3.836
100 + EDTA and C.A	1.060±0.17	0.557±0.07	1.617±0.06	1.903	1.060±0.00	0.552±0.16	1.612±0.00	1.920

Chlorophyll content (mg g^{-1}fw)								
15 mM Kg^{-1}					**20 mM Kg^{-1}**			
Treatment	**Chlorophyll a**	**Chlorophyll b**	**Total chlorophyll**	**Chlorophyll a/b**	**Chlorophyll a**	**Chlorophyll b**	**Total chlorophyll**	**Chlorophyll a/b**
Control	4.090±0.19	3.553±0.00	7.643±0.00	1.151	4.090±0.19	3.553±0.00	7.643±0.34	1.151
Control 25	1.009±0.00	0.420±0.16	1.429±0.00	2.402	1.009±0.11	0.420±0.01	1.429±0.22	2.402
25+ EDTA	1.156±0.00	0.603±0.15	1.759±0.15	1.915	1.230±0.01	0.582±0.29	1.812±0.00	2.110
Control 50	0.886±0.00	0.199±0.00	1.085±0.17	4.448	0.886±0.08	0.199±0.00	1.085±0.00	4.448
50+ EDTA	1.140±0.12	0.756±0.00	1.896±0.00	1.507	1.137±0.11	0.771±0.07	1.908±0.17	1.475
Control 75	0.835±0.18	0.382±0.16	1.217±0.18	2.184	0.835±0.00	0.382±0.22	1.217±0.38	2.184
75+ EDTA	1.145±0.00	0.761±0.00	1.906±0.00	1.504	1.151±0.15	0.782±0.00	1.933±0.00	1.472
Control 100	0.729±0.16	0.223±0.15	0.952±0.16	3.260	0.729±0.18	0.223±0.21	0.952±0.28	3.260
100+ EDTA	1.141±0.00	0.687±0.10	1.828±0.15	1.661	1.146±0.10	0.725±0.02	1.871±0.19	1.580
Control 25	0.713±0.18	0.402±0.16	1.115±0.00	1.773	0.713±0.00	0.402±0.02	1.115±0.00	1.773
25 + C.A	1.156±0.00	0.528±0.15	1.684±0.00	2.189	1.166±0.17	0.519±0.09	1.685±0.18	2.245
Control 50	0.849±0.00	0.263±0.00	1.112±0.66	3.228	0.849±0.00	0.263±0.00	1.112±0.19	3.220

50+ C.A	1.076±0.15	0.526±0.00	1.602±0.55	2.043	1.076±0.00	0.521±0.09	1.597±0.18	2.066
Control 75	0.672±0.17	0.214±0.16	0.886±0.00	3.131	0.672±0.08	0.214±0.08	0.886±0.20	3.131
75+ C.A	1.072±0.18	0.512±0.00	1.584±0.72	2.092	1.078±0.00	0.504±0.00	1.582±0.19	2.137
Control 100	0.656±0.16	0.171±0.16	0.827±0.00	3.840	0.656±0.05	0.171±0.04	0.827±0.00	3.840
100+ C.A	1.057±0.15	0.494±0.15	1.551±0.29	2.140	1.066±0.00	0.498±0.00	1.564±0.00	2.140
Control 25	0.713±0.00	0.402±0.16	1.115±0.17	1.773	0.713±0.20	0.402±0.00	1.115±0.00	1.773
25 + EDTA and C.A	1.040±0.00	0.458±0.15	1.498±0.00	2.270	0.288±0.00	1.620±0.0	1.908±0.10	1.770
Control 50	0.849±0.00	0.263±0.00	1.112±0.00	3.220	0.849±0.11	0.263±0.19	1.112±0.15	3.220
50 + EDTA and C.A	1.070±0.18	0.512±0.06	1.582±0.19	2.09	1.070±0.00	0.504±0.00	1.574±0.00	2.130
Control 75	0.672±0.00	0.214±0.00	0.886±0.00	3.131	0.672±0.19	0.214±0.16	0.886±0.11	3.131
75 + EDTA and C.A	1.140±0.17	0.761±0.22	1.901±0.16	1.500	1.150±0.00	0.782±0.00	1.932±0.19	1.470
Control 100	0.656±0.10	0.171±0.00	0.827±0.00	3.840	0.656±0.15	0.171±0.15	0.827±0.00	3.840
100 + EDTA and C.A	1.150±0.00	0.528±0.11	1.678±0.15	2.180	1.160±0.00	0.519±0.10	1.679±0.00	2.235

*Individual readings n =3 for each plant were averaged and are presented with ±SD.

*Significant correlation was found between chlorophyll activities at 0.05 level of significance; EDTA: Ethylene diamine tetraacetic acid; CA: Citric acid.

Table 4.16.2 Chlorophyll content of *P. juliflora* seedlings grown in soil for 60 days in different concentrations of F viz. control, 25, 50, 75 and 100 mg kg^{-1} NaF soil after treatment with chelates EDTA and CA

Chlorophyll content (mg g^{-1}fw)								
5 mM Kg^{-1}					**10 mM Kg^{-1}**			
Treatment	**Chlorophyll a**	**Chlorophyll b**	**Total chlorophyll**	**Chlorophyll a/b**	**Chlorophyll a**	**Chlorophyll b**	**Total chlorophyll**	**Chlorophyll a/b**
Control	4.304±0.00	3.762±0.37	8.066±0.70	1.144	4.304±0.00	3.762±1.00	8.066±0.99	1.144
Control 25	1.033±0.45	0.612±0.00	1.645±0.60	1.687	1.033±0.10	0.612±0.07	1.645±0.77	1.687

25+ EDTA	1.275±0.60	0.608±0.04	1.883±0.00	2.097	1.367±0.54	0.581±0.07	1.948±0.46	2.352
Control 50	0.916±0.00	0.388±0.06	1.304±0.29	2.360	0.916±0.05	0.388±0.09	1.304±0.19	2.360
50 + EDTA	1.302±0.11	0.776±0.07	2.078±0.67	1.677	1.306±0.00	0.773±0.08	2.079±0.76	1.689
Control 75	0.939±0.13	0.400±0.00	1.339±0.00	2.347	0.939±0.06	0.400±0.05	1.339±0.77	2.347
75+ EDTA	1.347±0.00	0.783±0.06	2.130±0.00	1.720	1.398±0.06	0.756±0.00	2.154±0.55	1.849
Control 100	0.930±0.23	0.398±0.06	1.328±0.11	2.336	0.930±0.07	0.398±0.07	1.328±0.45	2.336
100 + EDTA	1.310±0.81	0.672±0.00	1.982±0.00	1.949	1.317±0.07	0.687±0.06	2.004±0.71	1.917
Control 25	0.750±0.23	0.527±0.07	1.277±0.00	1.423	0.750±0.23	0.527±0.06	1.277±0.00	1.423
25+ C.A	1.207±0.00	0.633±0.05	1.841±0.56	1.906	0.952±0.25	0.732±0.07	1.684±0.44	1.300
Control 50	0.897±0.08	0.374±0.08	1.271±0.16	2.398	0.897±0.00	0.374±0.03	1.271±0.16	2.398
50+ C.A	1.135±0.81	0.660±0.00	1.795±0.00	1.719	1.144±0.06	0.661±0.08	1.805±0.19	1.730
Control 75	0.574±0.04	0.283±0.03	0.857±0.01	2.028	0.574±0.23	0.283±0.00	0.857±0.16	2.028
75+ C.A	1.782±0.00	0.644±0.00	2.426±0.00	2.767	1.178±0.19	0.649±0.00	1.827±0.17	1.815
Control 100	0.809±0.09	0.055±0.07	0.864±0.45	14.70	0.809±0.16	0.055±0.09	0.864±0.10	14.70
100+ C.A	0.853±0.00	0.763±0.08	1.616±0.00	1.117	0.858±0.17	0.767±0.05	1.625±0.08	1.118
Control 25	0.750±0.00	0.527±0.07	1.277±0.31	1.423	0.750±0.15	0.527±0.00	1.277±0.15	1.423
25 + EDTA and C.A	1.782±0.00	0.644±0.00	2.426±0.45	2.767	1.178±0.00	0.649±0.09	1.827±0.18	1.815
Control 50	0.897±0.18	0.374±0.10	1.271±0.00	2.398	0.897±0.00	0.374±0.08	1.271±0.17	2.398
50 + EDTA and C.A	1.135±0.10	0.660±0.06	1.795±0.15	1.719	1.144±0.17	0.661±0.06	1.805±0.00	1.730
Control 75	0.574±0.00	0.283±0.00	0.857±0.06	2.028	0.574±0.06	0.283±0.07	0.857±0.07	2.028
75 + EDTA and C.A	1.782±0.18	0.644±0.16	2.426±0.17	2.767	1.178±0.29	0.649±0.06	1.827±0.02	1.815
Control 100	0.809±0.00	0.055±0.00	0.864±0.00	14.70	0.809±0.00	0.055±0.00	0.864±0.09	14.70

100 + EDTA and C.A	0.853±0.00	0.763±0.11	1.616±0.03	1.117	0.858±0.15	0.767±0.17	1.625±0.00	1.118

Chlorophyll content (mg g^{-1}fw)								
15 mM Kg^{-1}					**20 mM Kg^{-1}**			
Treatment	**Chlorophyll a**	**Chlorophyll b**	**Total chlorophyll**	**Chlorophyll a/b**	**Chlorophyll a**	**Chlorophyll b**	**Total chlorophyll**	**Chlorophyll a/b**
Control	4.304±0.00	3.762±0.55	8.066±0.67	1.144	4.304±0.00	3.762±1.77	8.066±0.18	1.144
Control 25	1.033±0.88	0.612±0.00	1.645±0.00	1.687	1.033±0.19	0.612±0.00	1.645±0.55	1.687
25+ EDTA	1.372±0.56	0.603±0.09	1.975±0.89	2.275	1.405±0.17	0.656±0.04	2.061±0.00	2.141
Control 50	0.916±0.00	0.388±0.06	1.304±0.00	2.360	0.916±0.11	0.388±0.03	1.304±0.22	2.360
50+ EDTA	1.311±0.45	0.758±0.08	2.069±0.00	1.729	1.310±0.00	0.791±0.27	2.101±0.00	1.656
Control 75	0.939±0.55	0.400±0.00	1.339±0.59	2.347	0.939±0.19	0.400±0.00	1.339±0.44	2.347
75+ EDTA	1.438±0.88	0.755±0.00	2.193±0.00	1.904	1.366±0.19	0.792±0.27	2.158±0.00	1.724
Control 100	0.930±0.00	0.398±0.07	1.328±0.00	2.336	0.930±0.10	0.398±0.00	1.328±0.55	2.336
100+ EDTA	1.316±0.56	0.689±0.00	2.005±0.36	1.910	1.361±0.00	0.695±0.08	2.056±0.00	1.958
Control 25	0.750±0.08	0.527±0.07	1.277±0.00	1.423	0.750±0.22	0.527±0.05	1.277±0.07	1.423
25+ C.A	1.078±0.00	0.770±0.12	1.848±0.00	1.4	1.138±0.06	0.756±0.00	1.894±0.05	1.505
Control 50	0.897±0.17	0.374±0.00	1.271±0.17	2.398	0.897±0.00	0.374±0.06	1.271±0.00	2.398
50+ C.A	1.176±0.00	0.734±0.16	1.910±0.00	1.602	0.710±0.22	0.914±0.03	1.624±0.34	0.776
Control 75	0.574±0.00	0.283±0.00	0.857±0.00	2.028	0.574±0.00	0.283±0.00	0.857±0.05	2.028
75+ C.A	1.210±0.56	0.721±0.15	1.931±0.78	1.678	1.228±0.06	0.723±0.18	1.951±0.00	1.698
Control 100	0.809±0.00	0.055±0.17	0.864±0.00	14.70	0.809±0.00	0.055±0.19	0.864±0.02	14.70
100+ C.A	0.895±0.00	0.837±0.00	1.732±0.40	1.069	0.898±0.12	0.845±0.00	1.743±0.06	1.062
Control 25	0.750±0.09	0.527±0.17	1.277±0.00	1.423	0.750±0.05	0.527±0.16	1.277±0.00	1.423
25 + EDTA and C.A	1.210±0.00	0.721±0.00	1.931±0.20	1.678	1.228±0.00	0.723±0.00	1.951±0.07	1.698
Control 50	0.897±0.09	0.374±0.00	1.271±0.00	2.398	0.897±0.06	0.374±0.00	1.271±0.05	2.398

50 + EDTA and C.A	1.176±0.34	0.734±0.06	1.91±0.00	1.602	0.710±0.00	0.914±0.05	1.624±0.00	0.776
Control 75	0.574±0.00	0.283±0.00	0.857±0.00	2.028	0.574±0.06	0.283±0.04	0.857±0.04	2.028
75 + EDTA and C.A	1.210±0.00	0.721±0.06	1.931±0.07	1.678	1.228±0.00	0.723±0.00	1.951±0.00	1.698
Control 100	0.809±0.00	0.055±0.00	0.864±0.06	14.70	0.809±0.00	0.055±0.04	0.864±0.06	14.70
100 + EDTA and C.A	0.895±0.45	0.837±0.07	1.732±0.44	1.069	0.898±0.05	0.845±0.02	1.743±0.08	1.062

*Individual readings n =3 for each plant were averaged and are presented with ±SD.

*Significant correlation was found between chlorophyll activities at 0.05 level of significance;

EDTA: EthyleneEthylene diamine tetraacetic acid; CA: Citric acid.

Table 4.16.3: Chlorophyll content of *P. juliflora* seedlings grown in soil for 90 days in different concentrations of F viz. control, 25, 50, 75 and 100 mg kg^{-1} NaF soil after treatment with chelates EDTA and CA

Chlorophyll content (mg g^{-1}fw)								
5 mM Kg^{-1}					**10 mM Kg^{-1}**			
Treat-ment	**Chloro-phyll a**	**Chloro-phyll b**	**Total chlo-rophyll**	**Chlo-ro-phyll a/b**	**Chloro-phyll a**	**Chloro-phyll b**	**Total chlo-rophyll**	**Chlo-ro-phyll a/b**
Control	5.129±0.00	4.295±0.50	9.424±0.00	1.194	5.129±0.00	4.295±0.55	9.424±0.00	1.194
Control 25	1.425±0.00	0.691±0.00	2.116±0.55	2.062	1.425±0.44	0.691±0.00	2.116±0.45	2.062
25+ EDTA	2.425±0.55	0.338±0.07	2.763±0.00	7.174	2.501±0.56	0.328±0.20	2.829±0.00	7.625
Control 50	1.122±0.00	0.392±0.06	1.514±0.67	2.862	1.122±0.67	0.392±0.00	1.514±0.56	2.862
50+ EDTA	2.421±0.00	0.590±0.06	3.011±0.08	4.103	2.476±0.00	0.574±0.05	3.050±0.55	4.313
Control 75	1.195±0.00	0.427±0.00	1.622±0.08	2.798	1.195±0.78	0.427±0.00	1.622±0.15	2.798
75+ EDTA	2.526±0.55	0.535±0.07	3.061±0.00	4.721	2.533±0.45	0.556±0.06	3.089±0.00	4.555
Control 100	1.482±0.00	0.425±0.05	1.907±0.55	3.487	1.482±0.77	0.425±0.00	1.907±0.10	3.487
100+ EDTA	2.530±0.45	0.452±0.06	2.982±0.06	5.597	2.525±0.00	0.469±0.02	2.994±0.88	5.383

Control 25	0.889±0.07	0.508±0.00	1.397±0.04	1.75	0.889±0.00	0.508±0.00	1.397±0.10	1.75
25+ C.A	2.587±0.40	0.284±0.05	2.871±0.00	9.109	2.638±0.22	0.267±0.04	2.905±0.00	9.880
Control 50	1.041±0.66	0.360±0.00	1.401±0.78	2.891	1.041±0.44	0.360±0.00	1.401±0.45	2.891
50 + C.A	1.890±0.34	0.453±0.00	2.343±0.04	4.172	1.949±0.00	0.432±0.16	2.381±0.16	4.511
Control 75	0.732±0.00	0.307±0.06	1.039±0.04	2.384	0.732±0.09	0.307±0.01	1.039±0.00	2.384
75+ C.A	1.743±0.06	0.441±0.04	2.184±0.00	3.952	1.751±0.04	0.447±0.00	2.198±0.00	3.917
Control 100	0.960±0.07	0.157±0.00	1.117±0.07	6.114	0.960±0.00	0.157±0.08	1.117±0.56	6.114
100+ C.A	2.275±0.00	1.047±0.90	3.322±0.04	2.172	2.371±0.05	0.999±0.07	3.370±0.00	2.373
Control 25	0.889±0.05	0.508±0.00	1.397±0.00	1.75	0.889±0.06	0.508±0.00	1.397±0.08	1.75
25 + EDTA and C.A	1.932±0.04	0.461±0.06	2.393±0..06	4.190	2.146±0.00	0.525±0.08	2.671±0.16	4.087
Control 50	1.041±0.00	0.360±0.00	1.401±0.05	2.891	1.041±0.40	0.360±0.09	1.401±0.00	2.891
50 + EDTA and C.A	1.890±0.07	0.453±0.00	2.343±0.00	4.172	1.949±0.08	0.432±0.00	2.381±0.00	4.511
Control 75	0.732±0.45	0.307±0.00	1.039±0.34	2.384	0.732±0.00	0.307±0.10	1.039±0.15	2.384
75 + EDTA and C.A	1.743±0.00	0.441±0.05	2.184±0.00	3.952	1.751±0.44	0.447±0.00	2.198±0.00	3.917
Control 100	0.960±0.04	0.157±0.00	1.117±0.04	6.114	0.960±0.30	0.157±0.15	1.117±0.55	6.114
100 + EDTA and C.A	1.742±0.45	0.677±0.00	2.419±0.00	2.573	0.889±0.09	0.508±0.17	1.397±0.45	1.75

Chlorophyll content (mg g^{-1}fw)								
15 mM Kg^{-1}					**20 mM Kg^{-1}**			
Treatment	**Chlorophyll a**	**Chlorophyll b**	**Total chlorophyll**	**Chlorophyll a/b**	**Chlorophyll a**	**Chlorophyll b**	**Total chlorophyll**	**Chlorophyll a/b**
Control	5.129±0.50	4.295±0.78	9.424±1.66	1.194	5.129±0.08	4.295±1.55	9.424±0.78	1.194
Control 25	1.425±0.06	0.691±0.00	2.116±0.00	2.062	1.425±0.10	0.691±0.44	2.116±0.07	2.062
25+ EDTA	2.549±0.03	0.338±0.07	2.887±0.00	7.541	2.578±0.02	0.514±0.08	3.092±0.06	5.015

Control 50	1.122±0.00	0.392±0.05	1.514±0.06	2.862	1.122±0.30	0.392±0.05	1.514±0.06	2.862
50+ EDTA	2.524±0.17	0.554±0.18	3.078±0.07	4.555	2.566±0.08	0.540±0.05	3.106±0.04	4.751
Control	1.195±0.19	0.427±0.00	1.622±0.66	2.798	1.195±0.00	0.427±0.07	1.622±0.87	2.798
75+ EDTA	2.486±0.00	0.571±0.05	3.057±0.00	4.353	2.574±0.03	0.554±0.08	3.128±0.45	4.646
Control 100	1.482±0.10	0.425±0.08	1.907±0.55	3.487	1.482±0.04	0.425±0.06	1.907±0.67	3.487
100+ EDTA	2.568±0.70	0.526±0.00	3.094±0.30	4.882	2.526±0.06	0.574±0.03	3.100±0.04	4.400
Control	0.889±0.00	0.508±0.09	1.397±0.10	1.75	0.889±0.02	0.508±0.05	1.397±0.09	1.75
25+ C.A	3.001±0.17	0.355±0.08	3.356±0.66	8.453	2.999±0.10	0.378±0.06	3.377±0.79	7.933
Control 50	1.041±0.69	0.360±0.00	1.401±0.85	2.891	1.041±0.22	0.360±0.05	1.401±0.56	2.891
50+ C.A	2.146±0.00	0.525±0.09	2.671±0.06	4.087	1.845±0.04	1.371±0.06	3.216±0.03	1.345
Control 75	0.732±0.67	0.307±0.05	1.039±0.07	2.384	0.732±0.05	0.307±0.09	1.039±0.04	2.384
75+ C.A	1.743±0.45	0.663±0.00	2.406±0.18	2.628	1.747±0.04	0.677±0.06	2.424±0.45	2.580
Control 100	0.960±0.09	0.157±0.09	1.117±0.46	6.114	0.960±0.03	0.157±0.05	1.117±0.07	6.114
100+ C.A	2.406±0.00	1.117±0.50	3.523±0.00	2.153	2.501±0.04	1.107±0.04	3.608±0.57	2.259
Control 25	0.889±0.09	0.508±0.00	1.397±0.05	1.75	0.889±0.50	0.508±0.00	1.397±0.78	1.75
25+ EDTA and C.A	2.271±0.45	1.047±0.00	3.318±0.06	2.169	2.580±0.09	0.284±0.04	2.864±0.04	9.084
Control 50	1.041±0.00	0.360±0.06	1.401±0.07	2.891	1.041±0.44	0.360±0.03	1.401±0.06	2.891
50 + EDTA and C.A	2.521±0.67	0.165±0.00	2.686±0.05	15.27	2.631±0.05	0.856±0.06	3.487±0.03	3.073
Control 75	0.732±0.00	0.307±0.07	1.039±0.05	2.384	0.732±0.10	0.307±0.09	1.039±0.66	2.384
75 + EDTA and C.A	1.743±0.00	0.663±0.05	2.406±0.66	2.628	1.741±0.03	0.677±0.04	2.418±0.06	2.571
Control 100	0.960±0.05	0.157±0.07	1.117±0.30	6.114	0.960±0.44	0.157±0.05	1.117±0.07	6.114
100+ EDTA and C.A	2.372±0.00	0.999±0.00	3.371±0.50	2.374	2.581±0.09	0.282±0.08	2.863±0.09	9.152

*Individual readings n =3 for each plant were averaged and are presented with ±SD.

*Significant correlation was found between chlorophyll activities at 0.05 level of significance; EDTA: Ethylene diamine tetraacetic acid; CA: Citric acid.

Table 4.16.4: Chlorophyll content of *P. juliflora* seedlings grown in soil for 120 days in different concentrations of F viz. control, 25, 50, 75 and 100 mg kg^{-1} NaF soil after treatment with chelates EDTA and CA

Chlorophyll content (mg g^{-1}fw)								
5 mM Kg^{-1}					**10 mM Kg^{-1}**			
Treatment	**Chlorophyll a**	**Chlorophyll b**	**Total chlorophyll**	**Chlorophyll a/b**	**Chlorophyll a**	**Chlorophyll b**	**Total chlorophyll**	**Chlorophyll a/b**
Control	4.388±1.00	3.992±0.55	8.380±1.55	1.099	4.388±0.05	3.992±0.05	8.380±0.50	1.099
Control 25	1.235±0.05	0.613±0.00	1.848±0.00	2.014	1.235±0.08	0.613±0.00	1.848±0.00	2.014
25 + EDTA	1.772±0.05	0.493±0.00	2.265±0.00	3.594	1.951±0.50	0.457±0.06	2.408±0.09	4.269
Control 50	1.031±0.55	0.375±0.04	1.406±0.50	2.749	1.031±0.06	0.375±0.05	1.406±0.50	2.749
50 + EDTA	1.796±0.00	0.708±0.20	2.504±0.65	2.536	1.854±0.10	0.701±0.00	2.555±0.40	2.644
Control 75	1.170±0.00	0.367±0.00	1.537±0.45	3.188	1.170±0.07	0.367±0.00	1.537±0.09	3.188
75 + EDTA	1.905±0.45	0.683±0.00	2.588±0.00	2.789	1.917±0.55	0.687±0.10	2.604±0.05	2.790
Control 100	1.001±0.55	0.889±0.07	1.890±0.07	1.125	1.001±0.79	0.889±0.15	1.890±0.40	1.125
100 + EDTA	1.919±0.00	0.445±0.00	2.364±0.05	4.312	1.919±0.50	0.459±0.08	2.378±0.04	4.180
Control 25	0.911±0.06	0.480±0.05	1.391±0.08	1.897	0.911±0.40	0.480±0.00	1.391±0.09	1.897
25+ C.A	1.957±0.04	0.436±0.00	2.393±0.55	4.488	2.012±0.60	0.420±0.05	2.432±0.05	4.790
Control 50	1.066±0.08	0.323±0.00	1.389±0.09	3.300	1.066±0.50	0.323±0.08	1.389±0.04	3.300
50+ C.A	1.313±0.08	0.498±0.00	1.811±0.00	2.636	1.372±0.50	0.476±0.00	1.848±0.06	2.882
Control 75	0.648±0.00	0.289±0.06	0.937±0.00	2.242	0.648±0.07	0.289±0.04	0.937±0.05	2.242
75+ C.A	1.771±0.15	0.197±0.06	1.968±0.50	8.989	1.779±0.40	0.212±0.07	1.991±0.03	8.391
Control 100	0.965±0.15	0.064±0.00	1.029±0.40	15.07	0.965±0.00	0.064±0.00	1.029±0.09	15.07
100+ C.A	2.205±0.10	1.640±0.50	3.845±0.00	1.344	2.303±0.50	1.580±0.03	3.883±0.04	1.457
Control 25	0.911±0.00	0.480±0.00	1.391±0.00	1.897	0.911±0.40	0.480±0.15	1.391±0.10	1.897

25+ EDTA and C.A	1.022±0.55	0.506±0.00	1.528±0.54	2.019	0.600±0.50	0.328±0.06	0.928±0.08	1.829
Control 50	1.066±0.00	0.323±0.00	1.389±0.45	3.300	1.066±0.06	0.323±0.05	1.389±0.05	3.300
50+ EDTA and C.A	0.815±0.00	0.778±0.00	1.593±0.00	1.047	0.726±0.55	0.816±0.40	1.542±0.50	0.889
Control 75	0.648±0.00	0.289±0.05	0.937±0.06	2.242	0.648±0.05	0.289±0.00	0.937±0.05	2.242
75+ EDTA and C.A	0.979±0.05	0.539±0.00	1.518±0.06	1.816	0.989±0.00	0.534±0.00	1.523±0.00	1.852
Control 100	0.965±0.00	0.064±0.07	1.029±0.07	15.07	0.965±0.00	0.064±0.05	1.029±0.40	15.07
100+ EDTA and C.A	1.062±0.78	0.557±0.02	1.619±0.08	1.906	1.061±0.00	0.552±0.05	1.613±0.00	1.922

Chlorophyll content (mg g^{-1}fw)								
15 mM Kg^{-1}					**20 mM Kg^{-1}**			
Treatment	**Chlorophyll a**	**Chlorophyll b**	**Total chlorophyll**	**Chlorophyll a/b**	**Chlorophyll a**	**Chlorophyll b**	**Total chlorophyll**	**Chlorophyll a/b**
Control	4.388±0.50	3.992±0.00	8.380±0.40	1.099	4.388±0.00	3.992±0.05	8.380±1.00	1.099
Control 25	1.235±0.20	0.613±0.00	1.848±0.01	2.014	1.235±0.01	0.613±0.00	1.848±0.00	2.014
25+ EDTA	2.002±0.00	0.443±0.00	2.445±0.02	4.519	2.043±0.04	0.546±0.00	2.589±0.04	3.741
Control 50	1.031±0.40	0.375±0.00	1.406±0.05	2.749	1.031±0.00	0.375±0.05	1.406±0.00	2.749
50+ EDTA	1.905±0.03	0.676±0.40	2.581±0.01	2.818	1.949±0.55	0.659±0.42	2.608±0.00	2.957
Control 75	1.170±0.45	0.367±0.00	1.537±0.00	3.188	1.170±0.04	0.367±0.05	1.537±0.50	3.188
75+ EDTA	1.870±0.05	0.710±0.04	2.580±0.03	2.633	1.964±0.00	0.673±0.05	2.637±0.40	2.918
Control 100	1.001±0.60	0.889±0.04	1.890±0.05	1.125	1.001±0.40	0.889±0.10	1.890±0.40	1.125
100 + EDTA	1.956±0.65	0.594±0.05	2.550±0.06	3.292	1.914±0.30	0.646±0.04	2.560±0.05	2.962
Control 25	0.911±0.00	0.480±0.04	1.391±0.06	1.897	0.911±0.00	0.480±0.14	1.391±0.05	1.897
25+ C.A	1.964±0.00	0.590±0.06	2.554±0.02	3.328	1.963±0.00	0.606±0.02	2.569±0.04	3.239
Control 50	1.066±0.00	0.323±0.06	1.389±0.01	3.300	1.066±0.44	0.323±0.04	1.389±0.05	3.300
50+ C.A	1.592±0.50	0.438±0.05	2.030±0.04	3.634	1.206±0.04	2.023±0.02	3.229±0.04	0.596
Control 75	0.648±0.00	0.289±0.07	0.937±0.01	2.242	0.648±0.05	0.289±0.04	0.937±0.05	2.242

75 + C.A	1.780±0.50	0.347±0.08	2.127±0.04	5.129	1.785±0.06	0.347±0.00	2.132±0.04	5.144
Control 100	0.965±0.00	0.064±0.17	1.029±0.06	15.07	0.965±0.04	0.064±0.00	1.029±0.06	15.07
100+ C.A	2.357±0.60	1.532±0.06	3.889±0.05	1.538	2.452±0.05	1.518±0.04	3.970±0.05	1.615
Control 25	0.911±0.00	0.480±0.07	1.391±0.07	1.897	0.911±0.03	0.480±0.00	1.391±0.04	1.897
25+ EDTA and C.A	1.042±0.06	0.455±0.04	1.497±0.01	2.290	2.742±0.50	0.182±0.00	2.924±0.06	15.06
Control 50	1.066±0.00	0.323±0.05	1.389±0.04	3.300	1.066±0.04	0.323±0.04	1.389±0.05	3.300
50+ EDTA and C.A	0.819±0.40	0.874±0.02	1.693±0.05	0.937	0.853±0.40	0.763±0.00	1.616±0.20	1.117
Control 75	0.648±0.00	0.289±0.05	0.937±0.07	2.242	0.648±0.05	0.289±0.04	0.937±0.54	2.242
75+ EDTA and C.A	1.071±0.00	0.512±0.06	1.583±0.06	2.091	1.072±0.40	0.504±0.05	1.576±0.02	2.126
Control 100	0.965±0.00	0.064±0.00	1.029±0.03	15.07	0.965±0.06	0.064±0.05	1.029±0.06	15.07
100+ EDTA and C.A	1.152±0.00	0.528±0.08	1.680±0.08	2.181	1.163±0.42	0.519±0.05	1.682±0.69	2.240

*Individual readings n =3 for each plant were averaged and are presented with ±SD.

*Significant correlation was found between chlorophyll activities at 0.05 level of significance; EDTA: Ethylene diamine tetraacetic acid; CA: Citric acid.

4.13 EFFECT OF CHELATES (EDTA AND CA) ON ANTIOXIDANT ENZYME ACTIVITY

The effects of chelates (EDTA and CA) on antioxidant enzyme activities (POD, CAT and SOD) in leaves were found with different concentrations of F exposed to germplasm of *P. juliflora*. The antioxidant activity significantly increased with F concentration after 30 days up to 90 days and then decreased gradually after 120 days of treatment. The antioxidant enzyme activities were found significant at 20 mMEDTA of 1 kg soil at 100mgkg^{-1}NaF (0.160), (0.172) and (0.238) in treated plants of *P.juliflora* for 120 days as compared to CA and combined EDTA with CA (Figure 4.17.A to Figure 4.17.P). Heavy metals poisoning is the increased production of reactive oxygen species (ROS) due to intervention with electron transport activities, especially that of chloroplast membranes (Rocca et al., 2009). This increase in ROS exposes cells to oxidative stress leading to lipid peroxidation, biological macromolecule deterioration, membrane dismantling, ion leakage, and DNA-strand cleavage (Quartacci et al., 2001).

Plants route to a series of defense mechanisms that control uptake, accumulation and translocation of these dangerous elements and detoxify them. The maximum increase in POD, CAT and SOD activity was found at 90 days of exposure to 100 mg kg^{-1} NaF soil concentration and thereafter gradual decreases were observed after 90 days. The increased activities of POD, CAT and SOD in *P. juliflora* suggest that the plant depends on these antioxidative enzymes for the elimination of H_2O_2 under F stress. A further significant correlation was found between the increased POD, CAT and SOD activity with F ion concentration. The higher activity of antioxidant enzymes might be consequences of the strategies adopted by a plant for its survival imposed by F stress with the given treatment of EDTA, CA and combined EDTA+CA. Similar studies suggested that antioxidant activity was enhanced under the effect of Pb-EDTA to uptake mineral by *Phaseoulus vulgaris* plant (Geebelen *et al.*, 2002). However, less information is available about the relation of chelates and oxidative pathways of POD, SOD, and CAT, so it requires further research to understand the metabolic pathway generated under stress conditions.

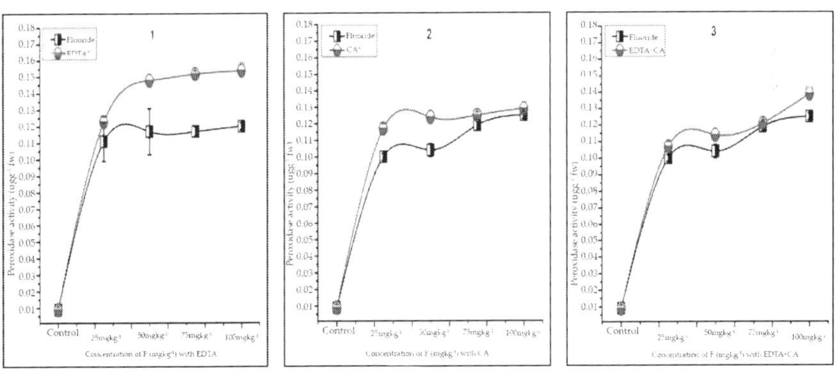

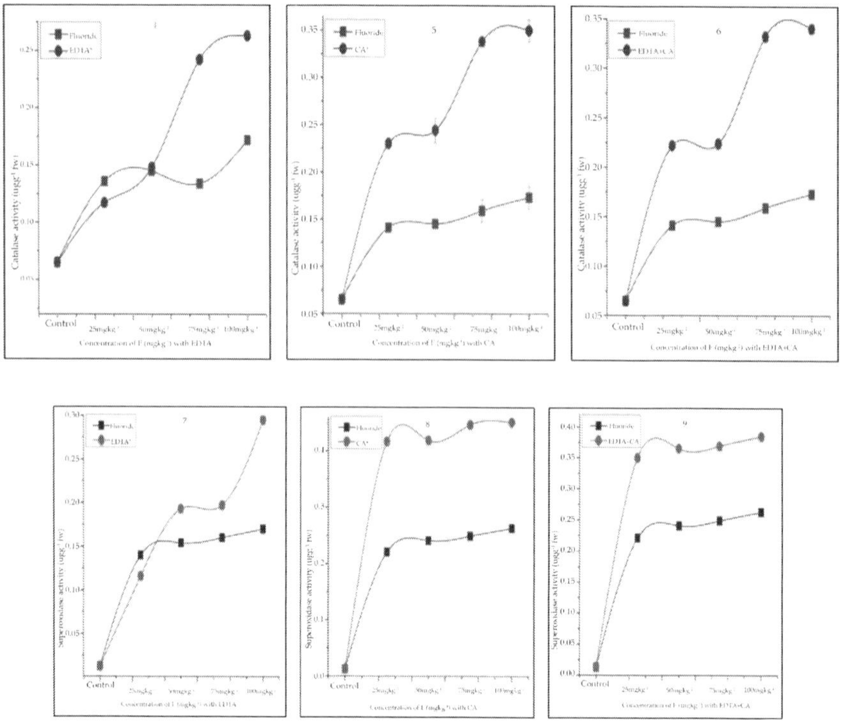

Figure 4.17.A: Effect of chelates (5 mM EDTA, CA and EDTA+CA) on antioxidant activity after 30 days of *P. juliflora* under F- stress with different concentrations of F viz. control, 25, 50, 75 & 100 mg kg^{-1} soil (1,2,3-peroxidase, 4,5,6-catalase and 7,8,9-superoxidaseEDTA-Ethylene diamine tetraacetic acid and CA-Citric acid where significantly indicated value*).

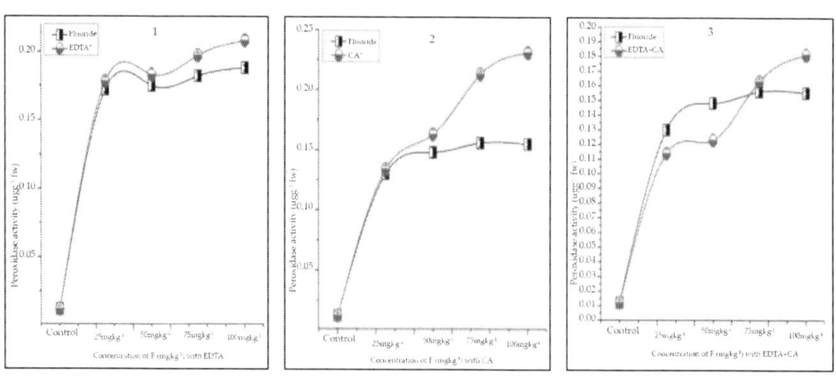

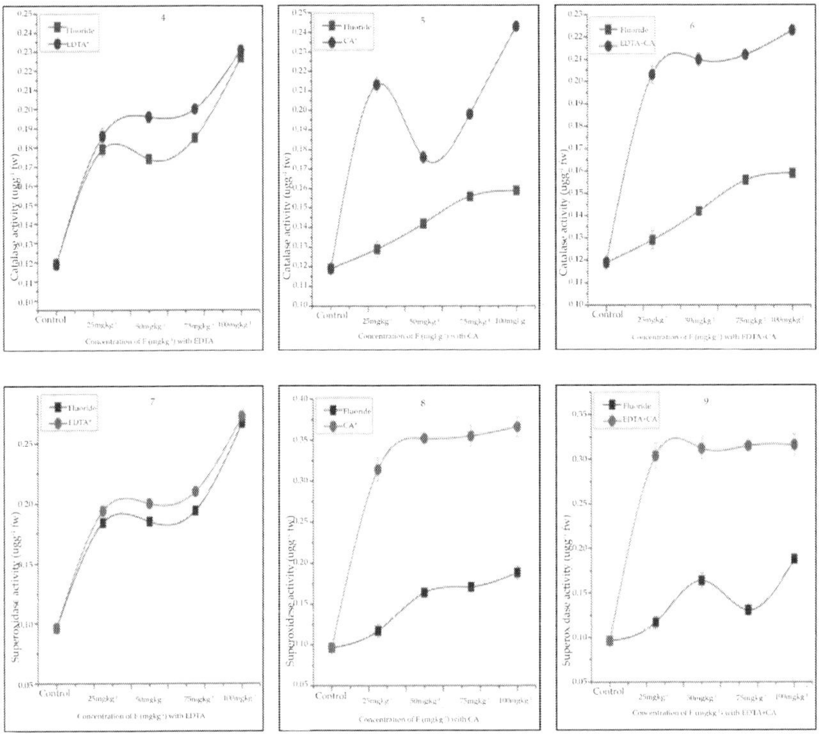

Figure 4.17.B: Effect of chelates (5 mM EDTA, CA and EDTA+CA) on antioxidant activity after 60 days of *P. juliflora* under F- stress with different concentrations of F viz. control, 25, 50, 75 & 100 mg kg^{-1} soil (1,2,3-peroxidase, 4,5,6-catalase and 7,8,9-superoxidaseEDTA-Ethylene diamine tetraacetic acid and CA-Citric acid where significantly indicated value*).

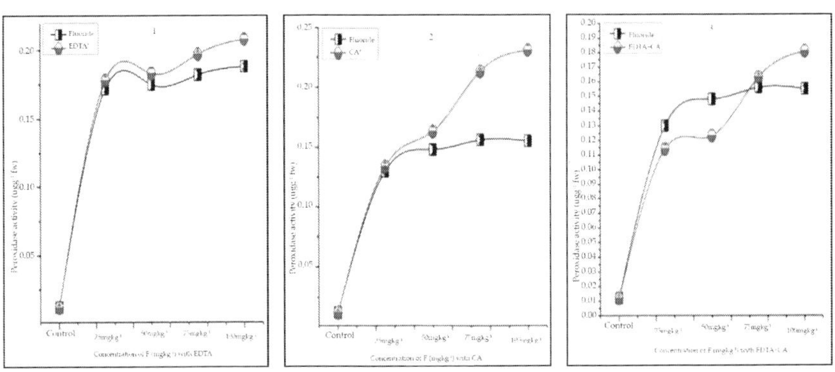

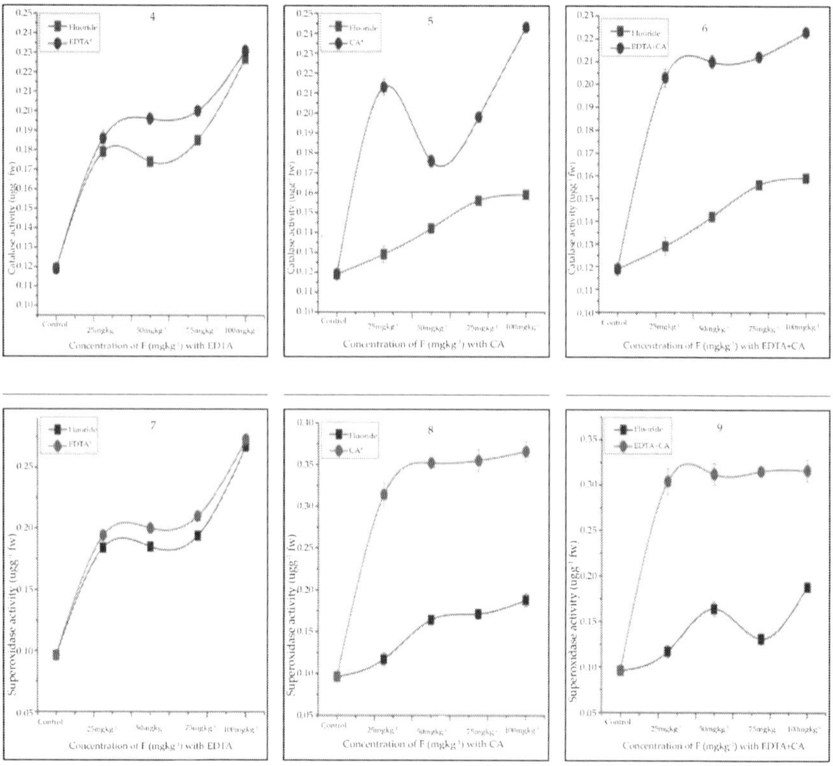

Figure 4.17.C: Effect of chelates (5 mM EDTA, CA and EDTA+CA) on anti-oxidant activity after 90 days of *P. juliflora* under F- stress with different concentrations of F viz. control, 25, 50, 75 & 100 mg kg^{-1} soil (1,2,3-peroxidase, 4,5,6-catalase and 7,8,9-superoxidaseEDTA-Ethylene diamine tetraacetic acid and CA-Citric acid where significantly indicated value*).

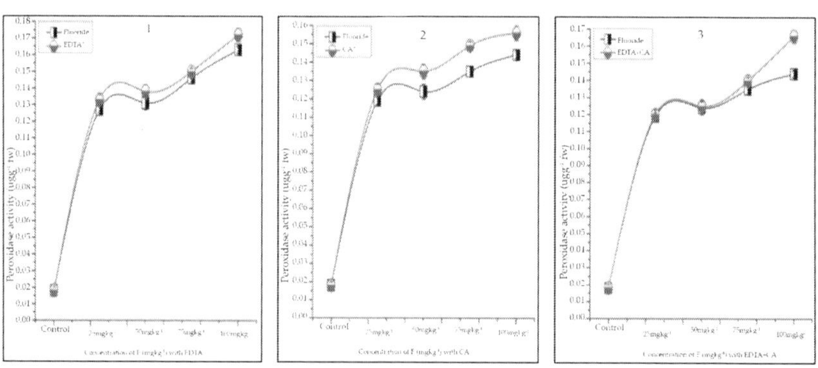

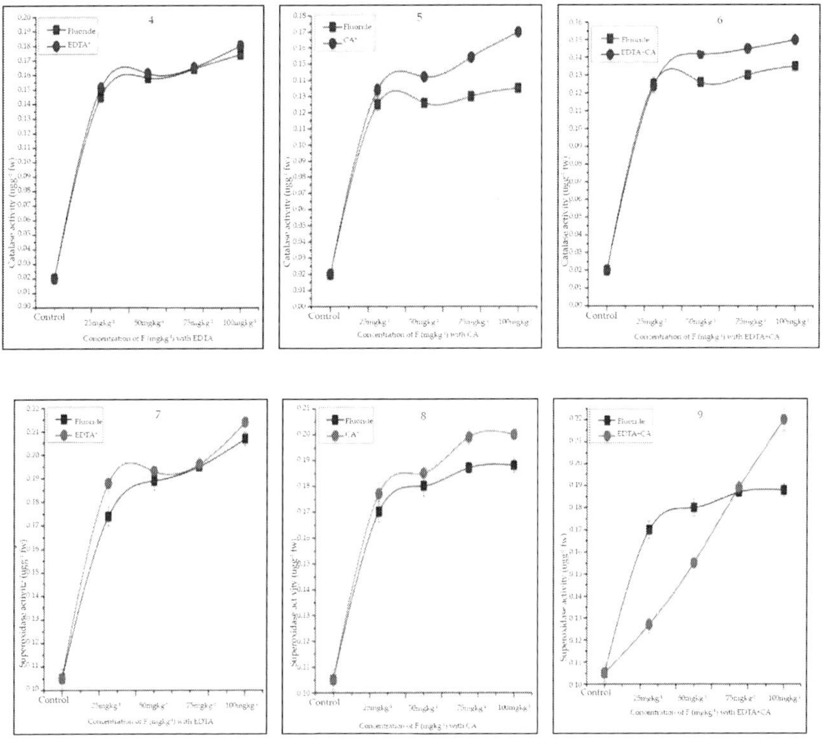

Figure 4.17.D: Effect of chelates (5 mM EDTA, CA and EDTA+CA) on anti-oxidant activity for 120 days of *P. juliflora* under F- stress with different concentrations of F viz. control, 25, 50, 75 & 100 mg kg^{-1} soil (1,2,3-peroxidase, 4,5,6-catalase and 7,8,9-superoxidaseEDTA-Ethylene diamine tetraacetic acid and CA-Citric acid where significantly indicated value*).

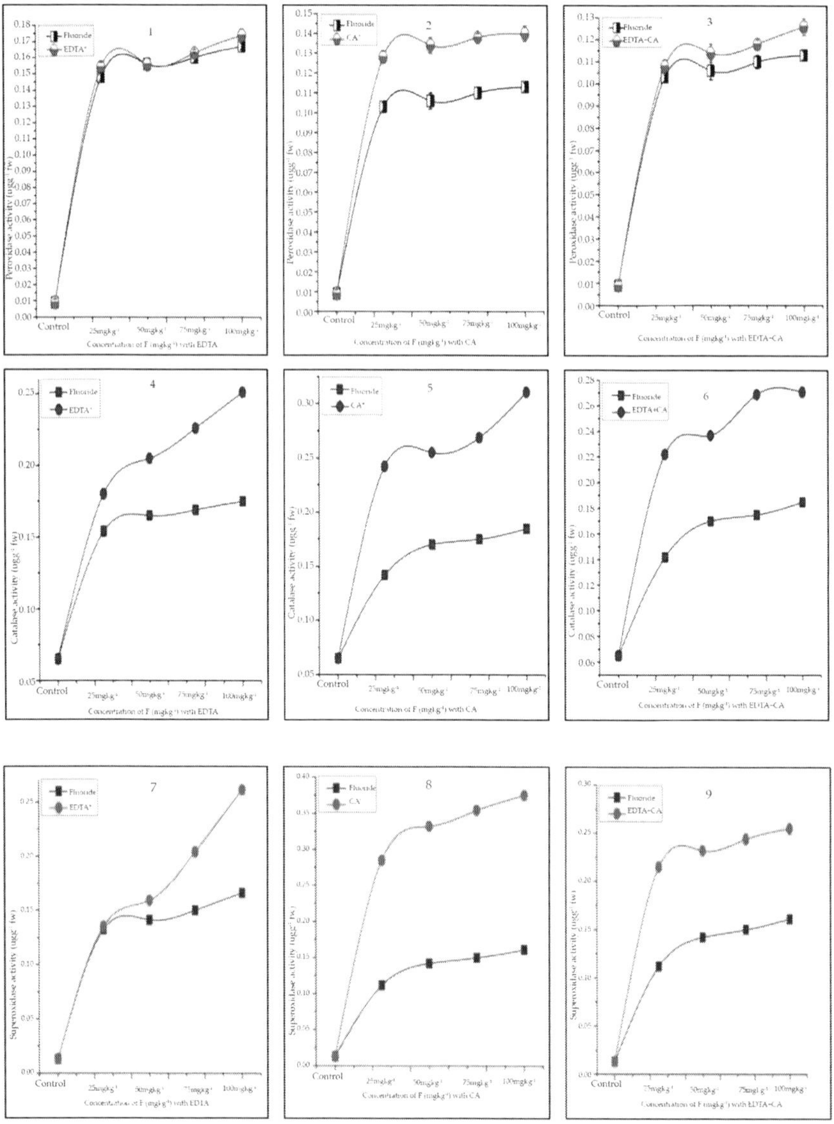

Figure 4.17.E: Effect of chelates (10 mM EDTA, CA and EDTA+CA) on antioxidant activity after 30 days of *P. juliflora* under F- stress with different concentrations of F viz. control, 25, 50, 75 & 100 mg kg^{-1} soil (1,2,3-peroxidase, 4,5,6-catalase and 7,8,9-superoxidaseEDTA-Ethylene diamine tetraacetic acid and CA-Citric acid where significantly indicated value*).

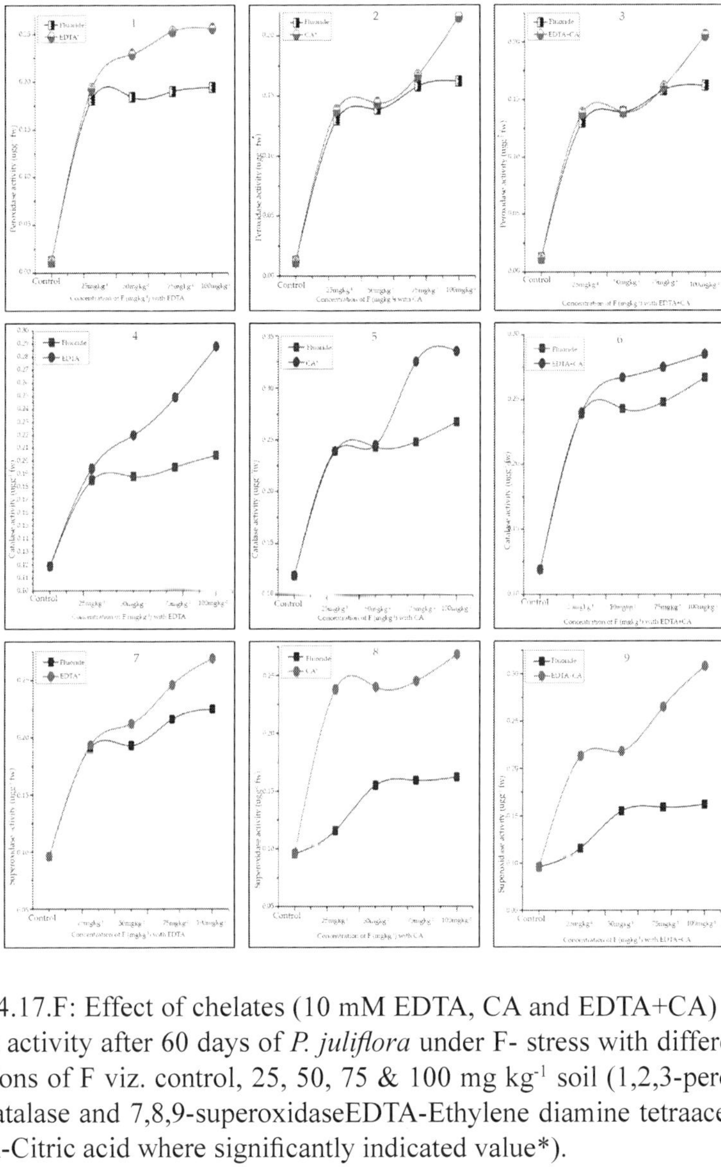

Figure 4.17.F: Effect of chelates (10 mM EDTA, CA and EDTA+CA) on antioxidant activity after 60 days of *P. juliflora* under F- stress with different concentrations of F viz. control, 25, 50, 75 & 100 mg kg^{-1} soil (1,2,3-peroxidase, 4,5,6-catalase and 7,8,9-superoxidaseEDTA-Ethylene diamine tetraacetic acid and CA-Citric acid where significantly indicated value*).

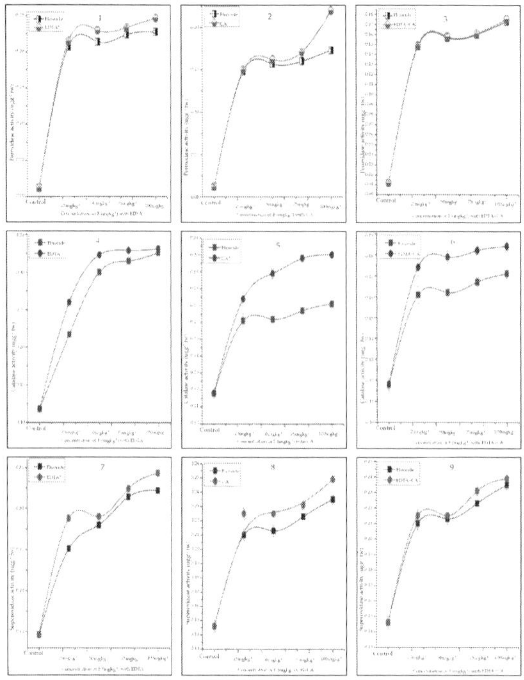

Figure 4.17.G: Effect of chelates (10 mM EDTA, CA and EDTA+CA) on antioxidant activity after 90 days of *P. juliflora* under F- stress with different concentrations of F viz. control, 25, 50, 75 & 100 mg kg^{-1} soil (1,2,3-peroxidase, 4,5,6-catalase and 7,8,9-superoxidase EDTA-Ethylene diamine tetraacetic acid and CA-Citric acid where significantly indicated value*).

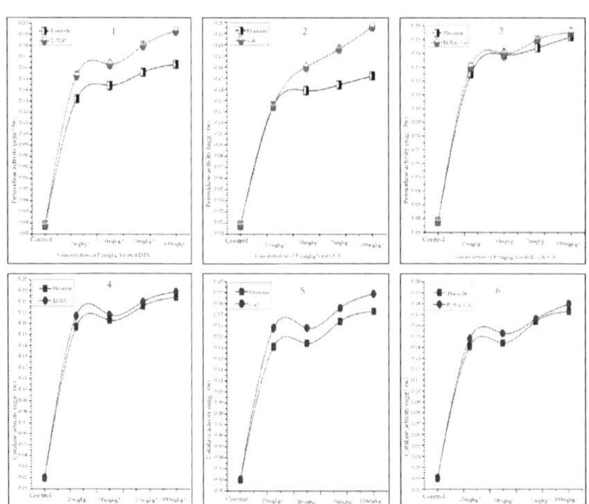

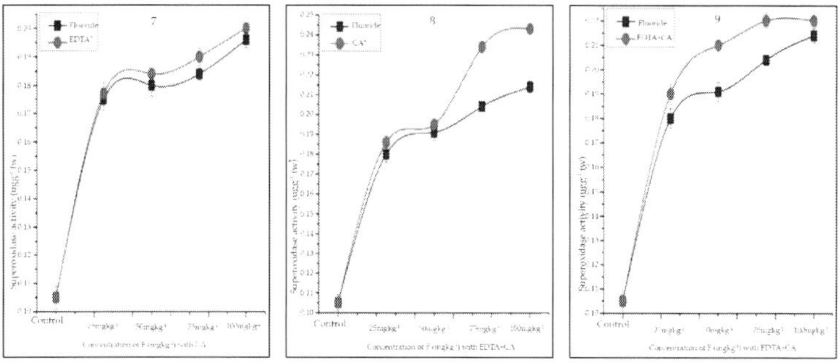

Figure 4.17.H: Effect of chelates (10 mM EDTA, CA and EDTA+CA) on antioxidant activity for 120 days of *P. juliflora* under F- stress with different concentrations of F viz. control, 25, 50, 75 & 100 mg kg^{-1} soil (1,2,3-peroxidase, 4,5,6-catalase and 7,8,9-superoxidaseEDTA-Ethylene diamine tetraacetic acid and CA-Citric acid where significantly indicated value*).

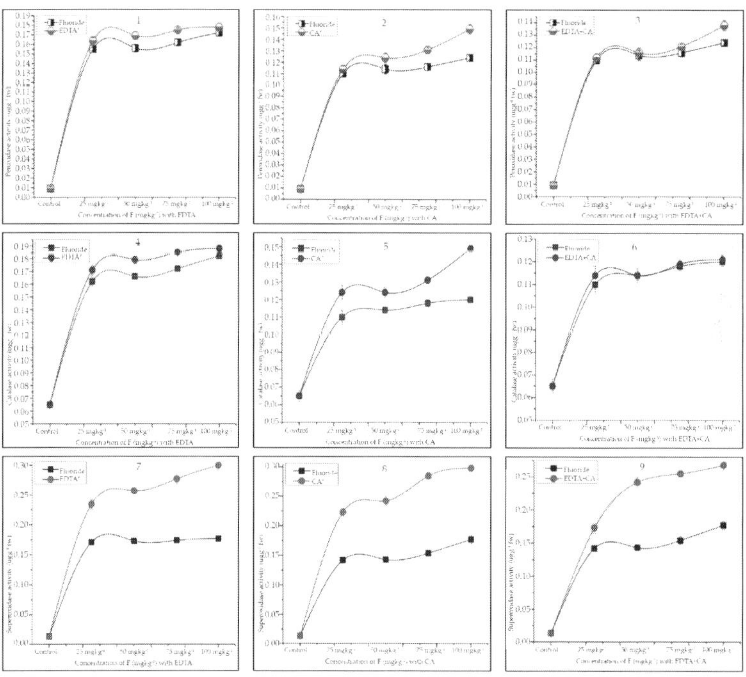

Figure 4.17.I: Effect of chelates (15 mM EDTA, CA and EDTA+CA) on antioxidant activity after 30 days of *P. juliflora* under F- stress with different concentrations of F viz. control, 25, 50, 75 & 100 mg kg^{-1} soil (1,2,3-peroxidase, 4,5,6-catalase and 7,8,9-superoxidaseEDTA-Ethylene diamine tetraacetic acid and CA-Citric acid where significantly indicated value*).

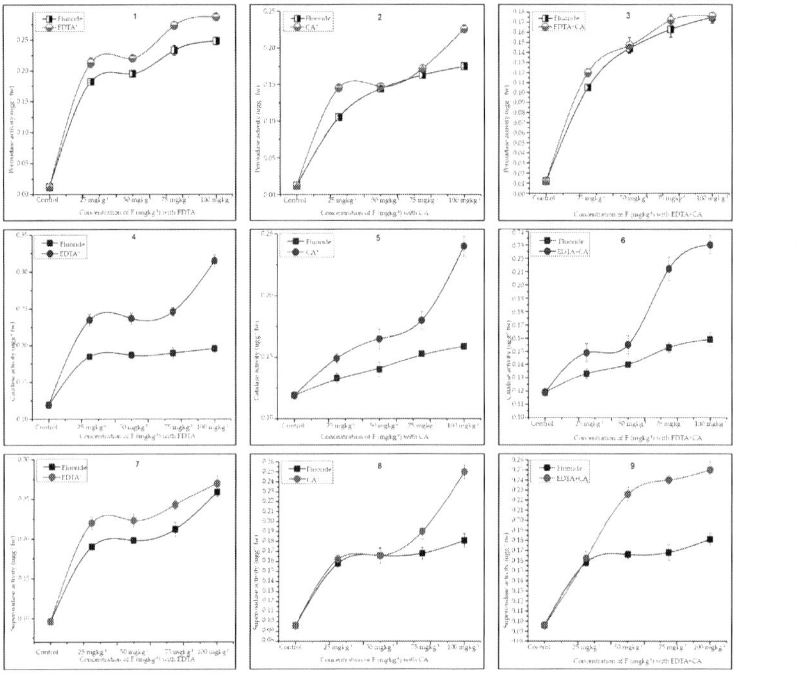

Figure 4.17.J: Effect of chelates (15 mM EDTA, CA and EDTA+CA) on anti-oxidant activity after 60 days of *P. juliflora* under F- stress with different concentrations of F viz. control, 25, 50, 75 & 100 mg kg^{-1} soil (1,2,3-peroxidase, 4,5,6-catalase and 7,8,9-superoxidaseEDTA-Ethylene diamine tetraacetic acid and CA-Citric acid where significantly indicated value*).

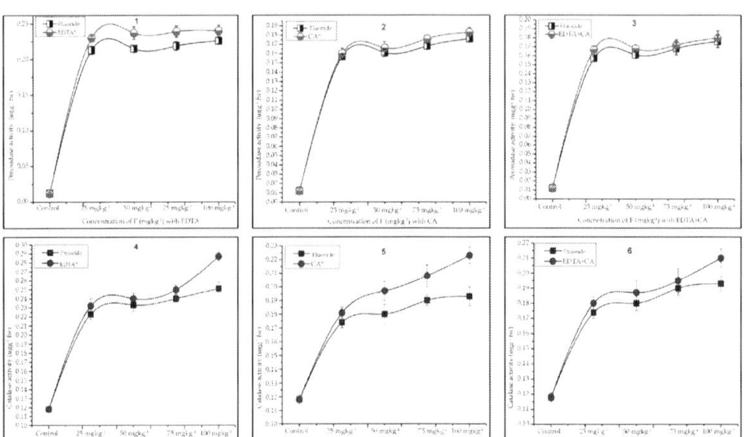

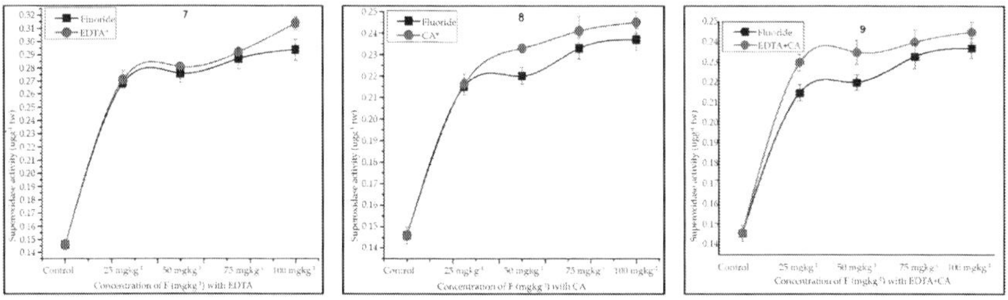

Figure 4.17.K: Effect of chelates (15 mM EDTA, CA and EDTA+CA) on antioxidant activity after 90 days of *P. juliflora* under F- stress with different concentrations of F viz. control, 25, 50, 75 & 100 mg kg^{-1} soil (1,2,3-peroxidase, 4,5,6-catalase and 7,8,9-superoxidaseEDTA-Ethylene diamine tetraacetic acid and CA-Citric acid where significantly indicated value*).

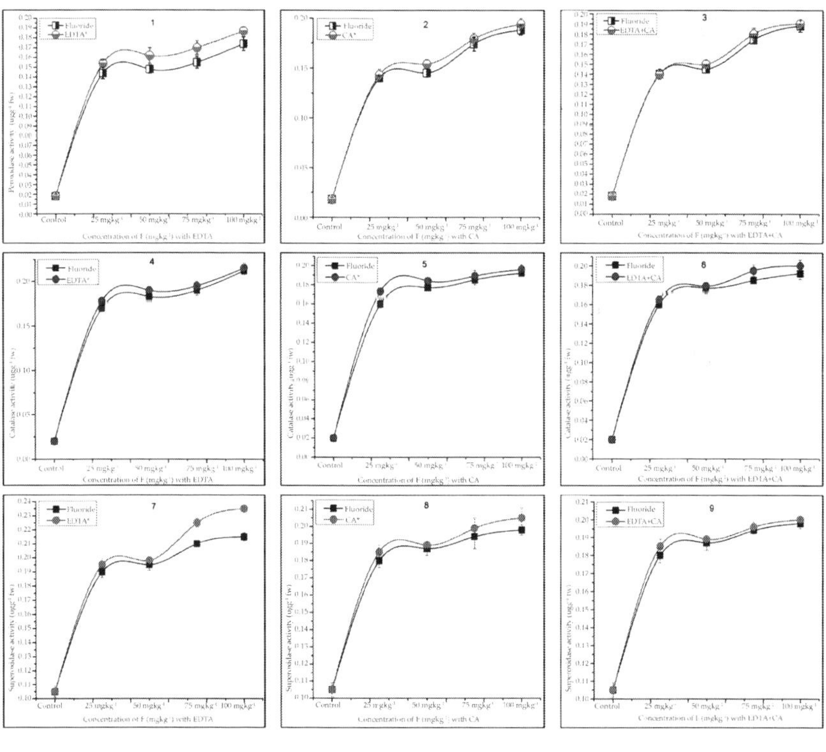

Figure 4.17.L: Effect of chelates (15 mM EDTA, CA and EDTA+CA) on antioxidant activity for 120 days of *P. juliflora* under F- stress with different concentrations of F viz. control, 25, 50, 75 & 100 mg kg^{-1} soil (1,2,3-peroxidase, 4,5,6-catalase and 7,8,9-superoxidaseEDTA-Ethylene diamine tetraacetic acid and CA-Citric acid where significantly indicated value*).

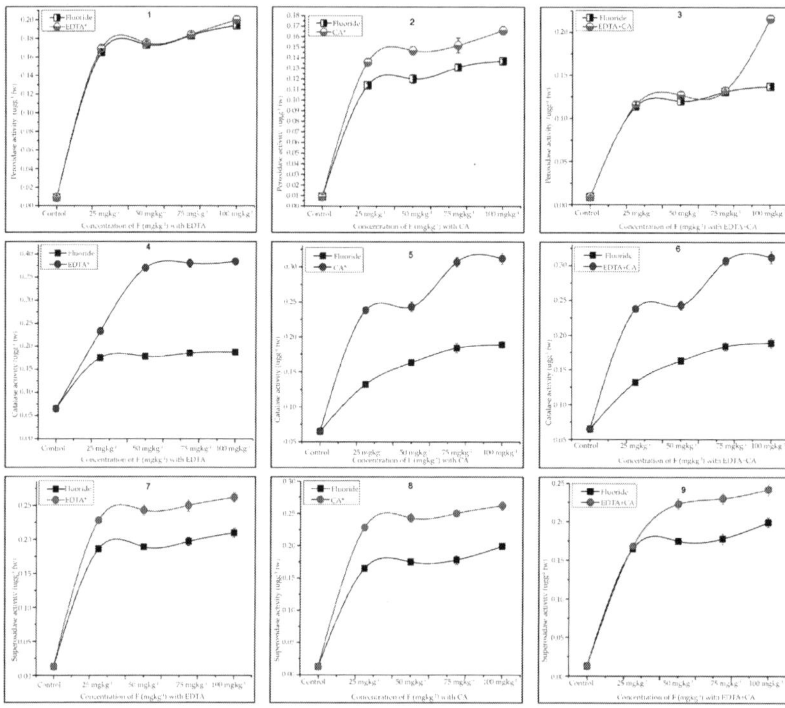

Figure 4.17.M: Effect of chelates (20 mM EDTA, CA and EDTA+CA) on antioxidant activity after 30 days of *P. juliflora* under F- stress with different concentrations of F viz. control, 25, 50, 75 & 100 mg kg^{-1} soil (1,2,3-peroxidase, 4,5,6-catalase and 7,8,9-superoxidaseEDTA-Ethylene diamine tetraacetic acid and CA-Citric acid where significantly indicated value*).

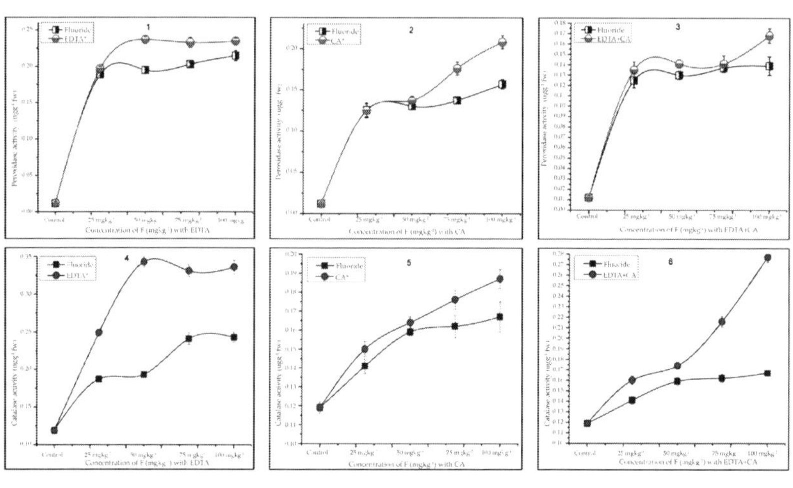

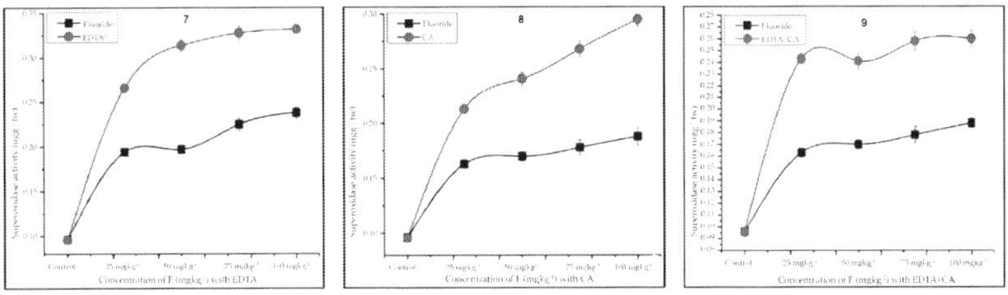

Figure 4.17.N: Effect of chelates (20 mM EDTA, CA and EDTA+CA) on antioxidant activity after 60 days of *P. juliflora* under F- stress with different concentrations of F viz. control, 25, 50, 75 & 100 mg kg^{-1} soil (1,2,3-peroxidase, 4,5,6-catalase and 7,8,9-superoxidaseEDTA-Ethylene diamine tetraacetic acid and CA-Citric acid where significantly indicated value*).

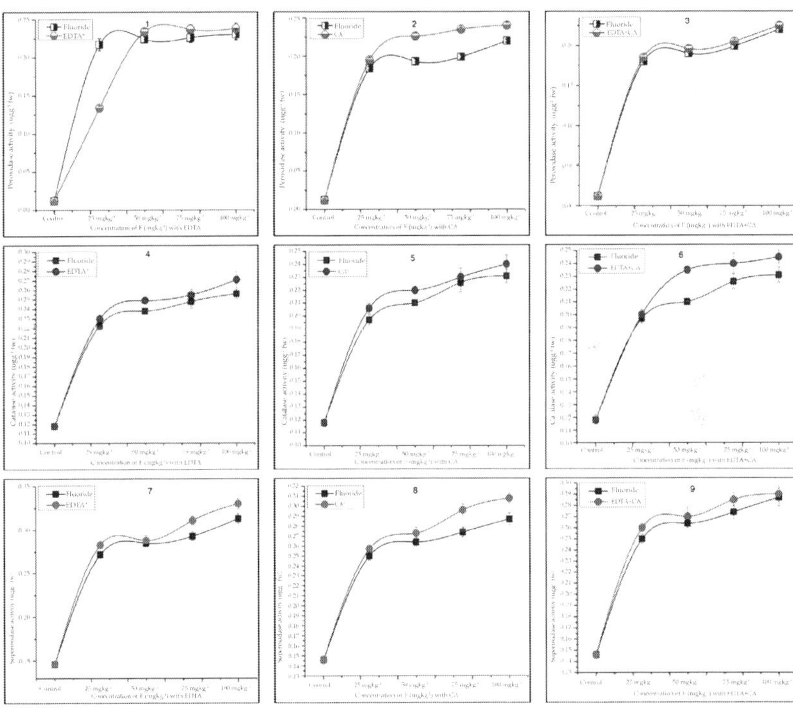

Figure 4.17.O: Effect of chelates (20 mM EDTA, CA and EDTA+CA) on antioxidant activity after 90 days of *P. juliflora* under F- stress with different concentrations of F viz. control, 25, 50, 75 & 100 mg kg^{-1} soil (1,2,3-peroxidase, 4,5,6-catalase and 7,8,9-superoxidaseEDTA-Ethylene diamine tetraacetic acid and CA-Citric acid where significantly indicated value*).

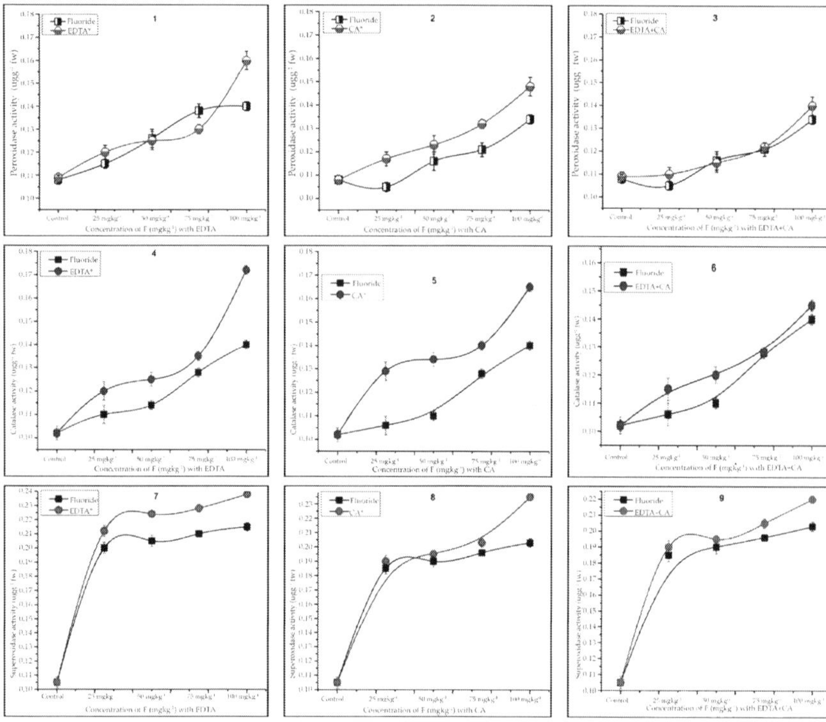

Figure 4.17.P: Effect of chelates (20 mM EDTA, CA and EDTA+CA) on antioxidant activity for 120 days of *P. juliflora* under F- stress with different concentrations of F viz. control, 25, 50, 75 & 100 mg kg^{-1} soil (1,2,3-peroxidase, 4,5,6-catalase and 7,8,9-superoxidaseEDTA-Ethylene diamine tetraacetic acid and CA-Citric acid where significantly indicated value*).

4.14 EFFECT OF CHELATES (EDTA AND CA) ON F ACCUMULATION

The present study was undertaken to check the F uptake capacity of *P. juliflora*. Hence the selected plants were grown in different concentrations of F viz. control, 25, 50, 75 and 100 mg kg^{-1} NaF soil with the treatment of chelates EDTA, CA and combined EDTA+CA. The total F accumulated by *P. juliflora* seedlings in the root (29.90 mg kg^{-1}), shoot (38.48 mg kg^{-1}) and remaining F is shown in (Figure 4. 18 and Table 4.17.1.2.3.4). It can be elucidated from the table (Table 4.17.1.2.3.4) that the total F content in root and shoot ranges between 18.56-29.90 and 3.56- 38.48 mg kg^{-1}to respectively, after given treatment with chelates. The highest F content was found in EDTA treated plants as compared to the control group. There was a significantly positive correlation found between the total F content in the root, shoot samples at 0.05 level of significance.

The maximum accumulation of F in root and shoot up to 120 days after given treatment are shown in (Table 4.18.1 to 4.18.4). Thus our results for F accumulation in shoots were higher than root at 120 days with chelates treatment.

It could be due to the soluble F fraction in the soil is taken up passively by roots as the ionic form at low pH which is more readily taken up by cell membranes and transport in shoot parts.

The efficiency of F accumulation in plant organs increases after given treatment with chelates in following order EDTA>CA>combined EDTA+CA.

The accumulation of F in the organs of plants depends on its concentration present in the soil, time of exposure and also their absorption capacity.The bioaccumulation factor (BF)ranged between (0.86-2.07) at 100 mg kg-1 NaF.*P. juliflora* showed a translocation factor (T.F) range from 0.98 to 1.28 at 100 mg kg-1 NaF (Figure 4.19 and Table 4.19.1.2.3.4).

Therefore, the T.F and B.Fwere found higher after 90 days with EDTA, CA and combined EDT+CA treatment.

The significant correlation between TF and BF was found at ($P < 0.05$) with the increase in F concentration.

The result of the present study shows that *P. juliflora* has a high capacity to accumulate more F which is available in the soil after the given treatment with EDTA, CA and combined EDTA+CA.

These studies suggested that tolerance of *P. juliflora* plants to F is a result of restrictive F transfer from the root to the aboveground parts. Fluoride localization in root cells indicated F is likely to be combined with sulfur-rich peptides or other constituents of the cell sap in the vacuole of the root (Dixit *et al.*, 2015).

Remarkable results were found from the present study, where the remaining F contents in the soil at 25 to 100 mg kg^{-1} NaF were reduced after 30 days.

This study has also confirmed by the growth and accumulation performance of *P. juliflora* plants under different F concentrations in soil with treated EDTA.

Table 4.17.1: Fluoride accumulation in the root of *P. juliflora* seedlings were grown in soil for 30 and 60 days in different concentrations of F viz. control, 25, 50, 75 and 100 mg kg^{-1} NaF after treatment with chelates EDTA and CA

Fluoride content (mg kg^{-1})									
30 days						**60 days**			
Treatment	**5 mM**	**10 mM**	**15 mM**	**20 mM**		**5 mM**	**10 mM**	**15 mM**	**20 mM**
Control	0.07±0.00	0.07±0.00	0.07±0.00	0.07±0.00		0.12±0.01	0.12±0.01	0.12±0.01	0.12±0.01
Control 25	2.06±0.11	2.06±0.11	2.06±0.11	2.06±0.11		2.43±0.40	2.43±0.40	2.43±0.40	2.43±0.40
25 + EDTA	3.23±0.37	3.43±0.37	3.56±0.49	3.63±0.55		4.40±0.52	4.53±0.61	4.66±0.70	4.80±0.80
Control 50	3.23±0.25	3.23±0.25	3.23±0.25	3.23±0.25		3.30±0.26	3.30±0.26	3.30±0.26	3.30±0.26
50 + EDTA	11.16±0.15	11.30±0.26	11.43±0.37	11.56±0.49		13.50±0.4	14.5±0.45	14.66±0.41	14.96±0.40
Control 75	4.23±0.25	4.23±0.25	4.23±0.25	4.23±0.25		4.26±0.25	4.26±0.25	4.26±0.25	4.26±0.25
75+ EDTA	15.16±0.15	15.3±0.26	15.43±0.37	15.56±0.49		19.40±0.5	19.9±1.22	20.3±1.83	21.10±2.97
Control 100	12.66±0.15	12.66±0.15	12.66±0.15	12.66±0.15		7.23±0.58	7.23±0.58	7.23±0.58	7.23±0.58
100 + EDTA	18.16±0.15	18.3±0.26	18.43±0.37	18.56±0.49		21.73±0.81	21.86±0.9	22.03±0.92	23.10±1.82
Control 25	2.16±0.15	2.16±0.15	2.16±0.15	2.16±0.15		2.36±0.35	2.36±0.35	2.36±0.35	2.36±0.35
25 + C.A	2.16±0.15	2.16±0.15	2.43±0.37	2.56±0.49		2.16±0.15	2.30±0.26	2.53±0.05	2.63±0.05
Control 50	2.53±0.10	2.53±0.10	2.53±0.10	2.53±0.10		2.93±2.16	2.93±2.16	2.93±2.16	2.93±2.16
50 + C.A	6.16±0.15	6.16±0.05	6.43±0.37	6.50±0.49		9.16±0.15	9.30±0.26	9.36±0.40	9.43±0.45
Control 75	5.33±0.35	5.33±0.35	5.33±0.35	5.33±0.35		6.36±0.35	6.36±0.35	6.36±0.35	6.36±0.35
75+ C.A	9.16±0.15	9.16±0.15	9.43±0.37	9.56±0.49		12.16±0.15	12.30±0.26	5.56±0.51	5.73±0.66
Control 100	2.56±0.11	2.56±0.11	2.56±0.11	2.56±0.11		7.36±0.35	7.36±0.35	7.36±0.35	7.36±0.35
100 + C.A	15.16±0.15	15.1±0.15	5.43±0.37	15.56±0.49		19.16±0.15	19.30±0.26	13.70±0.09	13.80±1.08
Control 25	1.33±0.30	1.33±0.30	1.33±0.30	1.33±0.30		1.40±0.36	1.40±0.36	1.40±0.36	1.740±0.36
25 + EDTA and C.A	1.76±0.11	1.73±0.23	1.80±0.20	1.93±0.05		2.10±0.10	2.4±0.10	2.53±0.05	2.63±0.05

Control 50	4.60±0.10	4.60±0.10	4.60±0.10	4.60±0.10		5.36±0.35	5.36±0.35	5.36±0.35	5.36±0.35
50 + EDTA and C.A	6.16±0.15	6.30±0.26	6.43±0.37	6.56±0.49		9.16±0.20	9.26±0.39	9.36±0.40	9.43±0.45
Control 75	5.63±0.25	5.63±0.25	5.63±0.25	5.63±0.25		5.53±0.47	6.53±0.47	6.53±0.47	6.53±0.47
75+ EDTA and C.A	4.16±0.15	4.30±0.26	4.43±0.37	4.56±0.49		5.33±0.30	5.46±0.45	5.56±0.51	5.73±0.66
Control 100	9.46±0.30	9.46±0.30	9.46±0.30	9.46±0.30		11.40±0.40	11.40±0.40	11.40±0.40	11.40±0.40
100+ EDTA and C.A	11.16±0.15	11.30±0.26	11.43±0.37	11.56±0.49		12.96±0.87	13.5±0.88	13.70±0.90	13.80±1.08

*Individual readings n =3 for each plant were averaged and are presented with ±SD.

*Significant correlation was found between Fluoride content at 0.05 level of significance;

EDTA: Ethylene diamine tetraacetic acid; CA: Citric acid.

Table 4.17.2: Fluoride accumulation in the root of *P. juliflora* seedlings were grown in soil for 90 and 120 days in different concentrations of F viz. control, 25, 50, 75 and 100 mg kg^{-1} NaF after treatment with chelates EDTA and CA

Fluoride content (mg kg^{-1})								
90 days					**120 days**			
Treatment	**5 mM**	**10 mM**	**15 mM**	**20 mM**	**5 mM**	**10 mM**	**15 mM**	**20 mM**
Control	0.13±0.03	0.13±0.03	0.13±0.03	0.13±0.03	3.18±0.16	3.18±0.16	3.18±0.16	3.18±0.00
Control 25	3.16±0.15	3.16±0.15	3.16±0.15	3.16±0.15	14.03±0.55	14.03±0.55	14.03±0.55	14.03±0.55
25+ EDTA	5.20±0.20	5.30±0.30	5.40±0.40	5.40±0.50	16.63±0.49	16.70±0.60	16.73±0.66	16.83±0.83
Control 50	4.46±0.05	4.46±0.05	4.46±0.05	4.46±0.05	18.36±0.30	18.36±0.30	18.36±0.30	18.36±0.30
50 + EDTA	16.20±0.2	16.26±0.25	16.33±0.30	16.40±0.3	18.46±0.56	18.70±0.75	18.80±0.85	18.90±0.90

Control 75	5.20±0.23	5.20±0.23	5.20±0.23	5.20±0.23	23.36±0.86	23.36±0.86	23.36±0.86	23.36±0.86
75+ EDTA	22.5±0.62	22.60±0.72	22.70±0.75	22.83±0.97	21.44±0.68	21.07±0.70	21.07±0.80	21.11±0.85
Control 100	22.36±0.35	22.36±0.35	22.36±0.35	22.36±0.35	26.36±0.15	26.36±0.15	26.36±0.15	26.36±0.15
100+ EDTA	25.26±0.30	25.33±0.35	25.40±0.40	25.80±1.05	29.46±0.50	29.63±0.70	29.70±0.75	29.90±1.01
Control 25	2.10±0.10	2.10±0.10	2.10±0.10	2.10±0.10	12.96±0.45	12.96±0.45	12.96±0.45	12.96±0.45
25 + C.A	2.53±0.41	2.60±0.34	2.73±0.25	2.83±0.15	13.26±0.30	13.36±0.35	13.46±0.45	13.53±0.50
Control 50	6.13±0.11	6.13±0.11	6.13±0.11	6.13±0.11	17.36±0.32	17.36±0.32	17.36±0.32	17.36±0.32
50 + C.A	11.13±1.10	11.26±1.20	11.36±1.30	11.46±1.30	12.23±0.25	12.36±0.35	12.43±0.40	12.56±0.51
Control 75	7.20±0.20	7.20±0.20	7.20±0.20	7.20±0.20	16.83±0.11	16.83±0.11	16.83±0.11	16.83±0.11
75 + C.A	18.50±0.62	18.66±0.7	18.76±0.75	18.90±0.90	23.40±0.52	23.50±0.61	23.60±0.65	23.70±0.75
Control 100	22.30±0.26	22.30±0.26	22.30±0.26	22.30±0.26	24.43±0.40	24.43±0.40	24.43±0.40	24.43±0.40
100 + C.A	24.33±0.30	24.43±0.40	24.53±0.47	24.60±0.52	27.40±0.45	27.46±0.50	27.66±0.70	27.83±0.80
Control 25	2.33±0.03	2.33±0.03	2.33±0.03	2.33±0.03	12.11±0.19	12.11±0.19	12.11±0.19	12.11±0.19
25+ EDTA and C.A	3.23±0.25	3.36±0.35	3.43±0.40	3.50±0.45	16.10±1.35	16.04±1.39	15.92±1.66	16.14±1.28
Control 50	7.33±0.30	7.33±0.30	7.33±0.30	7.33±0.30	13.29±0.25	13.29±0.25	13.29±0.25	13.29±0.25
50 + EDTA and C.A	12.33±0.30	12.46±0.45	12.53±0.50	12.76±0.75	22.88±0.50	23.25±0.22	23.29±0.25	23.29±0.25
Control 75	6.56±0.20	6.56±0.20	6.56±0.20	6.56±0.20	16.10±0.50	16.10±0.50	16.10±0.50	16.10±0.50
75 + EDTA and C.A	5.33±0.30	5.43±0.37	5.50±0.43	5.60±0.55	23.11±0.19	23.10±0.17	23.22±0.38	23.18±0.31
Control 100	17.33±0.30	17.33±0.30	17.33±0.30	17.33±0.30	18.51±0.44	18.51±0.34	18.51±0.40	18.51±0.40
100 + EDTA and C.A	18.40±0.36	18.50±0.45	18.60±0.52	18.75±0.66	27.22±0.38	27.22±0.38	27.22±0.38	27.22±0.38

*Individual readings n =3 for each plant were averaged and are presented with ±SD.

*Significant correlation was found between Fluoride content at 0.05 level of significance;

EDTA: Ethylene diamine tetraacetic acid; CA: Citric acid.

Table 4.17.3: Fluoride accumulation in the shoot of *P. juliflora* seedlings were grown in soil for 30 and 60 days in different concentrations of F viz. control, 25, 50, 75 and 100 mg kg^{-1} NaF after treatment with chelates EDTA and CA

Fluoride content (mg kg^{-1})								
30 days					**60 days**			
Treat-ment	**5 mM**	**10 mM**	**15 mM**	**20 mM**	**5 mM**	**10 mM**	**15 mM**	**20 mM**
Control	0.31±0.01	0.31±0.01	0.31±0.01	0.31±0.01	0.44±0.01	0.44±0.01	0.44±0.01	0.44±0.01
Control 25	1.20±0.20	1.20±0.20	1.20±0.20	1.20±0.20	1.50±0.00	1.50±0.00	1.50±0.00	1.50±0.00
25 + EDTA	1.81±0.01	1.83±0.02	1.84±0.03	1.85±0.04	2.20±0.17	2.30±0.26	2.43±0.37	2.56±0.49
Control 50	2.43±0.12	2.43±0.12	2.43±0.12	2.43±0.12	3.20±0.20	3.20±0.20	3.20±0.20	3.20±0.20
50 + EDTA	3.41±0.01	3.41±0.01	3.41±0.01	3.41±0.01	4.16±0.15	4.30±0.26	4.43±0.37	4.56±0.49
Control 75	1.22±0.12	1.22±0.12	1.22±0.12	1.22±0.12	1.46±0.11	1.46±0.11	1.46±0.11	1.46±0.11
75 + EDTA	5.62±0.02	5.62±0.02	5.62±0.02	5.62±0.02	6.16±0.15	6.30±0.26	6.43±0.37	6.56±0.49
Control 100	6.46±0.20	6.46±0.20	6.46±0.20	6.46±0.20	7.80±0.10	7.80±0.10	7.80±0.10	7.80±0.10
100 + EDTA	3.26±0.25	3.40±0.35	3.44±0.38	3.56±0.38	3.91±0.01	3.93±0.02	3.94±0.03	3.95±0.04
Control 25	1.42±0.36	1.42±0.36	1.42±0.36	1.42±0.36	1.52±0.02	1.52±0.02	1.52±0.02	1.52±0.02
25 + C.A	0.74±0.03	0.74±0.03	0.74±0.03	0.74±0.03	0.93±0.06	0.94±0.05	0.95±0.05	0.95±0.04
Control 50	4.60±0.26	4.60±0.26	4.60±0.26	4.60±0.26	2.38±0.03	2.38±0.03	2.38±0.03	2.38±0.03
50 + C.A	1.81±0.01	1.81±0.02	1.84±0.03	1.85±0.04	2.16±0.15	2.30±0.26	2.43±0.37	2.56±0.49
Control 75	3.73±0.15	3.73±0.15	3.73±0.15	3.73±0.15	2.74±0.07	2.74±0.07	2.74±0.07	2.74±0.07
75 + C.A	2.34±0.04	2.36±0.07	2.41±0.06	2.62±0.25	3.91±0.01	3.93±0.02	3.94±0.03	3.95±0.04
Control 100	7.56±0.30	7.56±0.30	7.56±0.30	7.56±0.30	4.29±0.05	4.29±0.05	4.29±0.05	4.29±0.05
100 + C.A	5.41±0.01	5.43±0.03	5.44±0.03	5.45±0.04	6.81±0.02	6.82±0.03	6.83±0.04	6.85±0.06
Control 25	0.98±0.00	0.98±0.00	0.98±0.00	0.98±0.00	1.06±0.05	1.06±0.05	1.06±0.05	1.06±0.05
25 + EDTA and C.A	1.45±0.10	1.55±0.09	1.56±0.09	1.60±0.07	2.13±0.15	2.20±0.20	2.33±0.35	2.40±0.40
Control 50	3.23±0.15	3.23±0.15	3.23±0.15	3.23±0.15	5.20±0.26	5.20±0.26	5.20±0.26	5.20±0.26

50 + EDTA and C.A	3.40±0.01	3.43±0.02	3.44±0.03	3.45±0.04	4.34±0.29	4.47±0.06	4.54±1.05	4.66±0.21
Control 75	3.30±0.26	3.30±0.26	3.30±0.26	3.30±0.26	4.20±0.21	4.20±0.21	4.20±0.21	4.20±0.21
75+ EDTA and C.A	3.71±0.01	3.73±0.02	3.74±0.03	3.75±0.04	4.13±0.15	4.30±0.26	4.43±0.37	4.56±0.49
Control 100	3.80±0.10	3.80±0.10	3.80±0.10	3.80±0.10	7.26±0.25	7.26±0.25	7.26±0.25	7.26±0.25
100+ EDTA and C.A	4.51±0.01	4.53±0.02	4.54±0.03	4.55±0.04	5.23±0.20	5.36±0.32	5.50±0.43	5.60±0.51

*Individual readings n =3 for each plant were averaged and are presented with ±SD.

*Significant correlation was found between Fluoride content at 0.05 level of significance; EDTA: Ethylene diamine tetraacetic acid; CA: Citric acid.

Table 4.17.4: Fluoride accumulation in the shoot of *P. juliflora* seedlings were grown in soil for 90 and 120 days in different concentrations of F viz. control, 25, 50, 75 and 100 mg kg^{-1} NaF after treatment with chelates EDTA and CA

Fluoride content (mg kg^{-1})								
90 days					**120 days**			
Treat-ment	**5 mM**	**10 mM**	**15 mM**	**20 mM**	**5 mM**	**10 mM**	**15 mM**	**20 mM**
Con-trol	0.55±0.00	0.55±0.00	0.55±0.00	0.55±0.00	2.67±0.00	2.67±0.00	2.67±0.00	2.67±0.00
Con-trol 25	1.76±0.05	1.76±0.05	1.76±0.05	1.76±0.05	12.10±0.15	12.10±0.15	12.10±0.15	12.10±0.15
25 + EDTA	3.16±0.15	3.30±0.26	3.43±0.37	3.56±0.49	7.83±0.85	7.83±0.85	7.82±0.83	7.86±0.89
Con-trol 50	3.66±0.11	3.66±0.11	3.66±0.11	3.66±0.11	16.25±0.44	16.25±0.44	16.25±0.44	16.25±0.44
50+ EDTA	5.16±0.15	5.30±0.26	5.43±0.37	5.56±0.49	19.07±0.12	19.14±0.25	19.11±0.11	19.07±0.07
Con-trol 75	1.23±0.15	1.23±0.15	1.23±0.15	1.23±0.15	21.11±0.19	21.11±0.19	22.11±0.19	22.11±0.19
75 + EDTA	7.16±0.15	7.30±0.26	7.43±0.37	7.56±0.49	22.11±1.64	22.25±1.51	22.42±1.37	22.47±1.35
Con-trol 100	9.73±0.05	9.73±0.05	9.73±0.05	9.73±0.05	25.96±1.94	25.96±1.94	26.96±1.94	27.96±1.94
100 + EDTA	4.16±0.15	4.30±0.26	4.43±0.37	4.56±0.49	36.43±0.67	36.50±0.79	37.73±0.40	38.48±0.02

Control 25	1.73±0.05	1.73±0.05	1.73±0.05	1.73±0.05	10.26±0.65	10.26±0.65	10.26±0.65	10.26±0.65
25+ C.A	1.31±0.01	1.33±0.02	1.34±0.03	1.35±0.04	11.30±0.26	11.43±0.37	11.53±0.47	11.74±0.72
Control 50	3.20±0.20	3.20±0.20	3.20±0.20	3.20±0.20	11.34±0.29	11.34±0.29	11.34±0.29	11.34±0.29
50 + C.A	3.16±0.15	3.30±0.26	3.43±0.37	3.56±0.49	12.25±0.22	12.29±0.27	12.37±0.34	12.40±0.38
Control 75	3.40±0.20	3.40±0.20	3.40±0.20	3.40±0.20	17.29±0.00	17.29±0.00	17.29±0.00	17.29±0.00
75+ C.A	4.16±0.15	4.30±0.26	4.43±0.37	4.56±0.49	22.92±0.93	23.18±0.83	23.48±0.90	24.14±0.79
Control 100	5.23±0.25	5.23±0.25	5.23±0.25	5.23±0.25	23.31±1.05	23.31±1.05	23.31±1.05	23.31±1.05
100 + C.A	7.16±0.15	7.30±0.26	7.43±0.37	7.56±0.49	28.11±0.19	28.22±0.38	28.25±1.09	28.50±1.40
Control 25	1.40±0.10	1.40±0.10	1.40±0.10	1.40±0.10	11.73±0.11	11.73±0.11	11.73±0.11	11.73±0.11
25 + EDTA and C.A	2.40±0.26	2.56±0.20	2.73±0.15	2.90±0.10	14.44±0.50	14.44±0.50	14.51±0.50	14.55±0.50
Control 50	6.26±0.25	6.26±0.25	6.26±0.25	6.26±0.25	12.40±0.26	12.40±0.26	12.40±0.26	12.40±0.26
50+ EDTA and C.A	5.16±0.15	5.30±0.26	5.43±0.37	5.56±0.49	16.18±0.31	16.25±0.44	16.26±0.45	16.29±0.51
Control 75	4.56±0.15	4.56±0.15	4.56±0.15	4.56±0.15	15.50±0.10	15.50±0.010	15.50±0.10	15.50±0.10
75 + EDTA and C.A	5.16±0.15	5.30±0.26	5.43±0.37	5.56±0.49	22.11±0.19	22.15±0.25	22.18±0.32	22.22±0.38
Control 100	9.63±0.15	9.63±0.15	9.63±0.15	9.63±0.15	17.42±0.36	17.42±0.36	17.42±0.36	17.42±0.36
100 + EDTA and C.A	6.16±0.15	6.30±0.26	6.43±0.37	6.56±0.49	26.15±0.02	26.22±0.38	26.26±0.45	26.62±0.54

*Individual readings n =3 for each plant were averaged and are presented with ±SD.

*Significant correlation was found between Fluoride content at 0.05 level of significance; EDTA: Ethylene diamine tetraacetic acid; CA: Citric acid.

Table 4.18.1: Remaining F concentration in the soil after harvesting the plantlets of *P. juliflora* for 30 and 60 days in different concentrations of F viz. control, 25, 50, 75, and 100 mg kg^{-1} NaF after treatment with chelates EDTA and CA

Fluoride content (mg kg^{-1})								
30 days					60 days			
Treatment	5 mM	10 mM	15 mM	20 mM	5 mM	10 mM	15 mM	20 mM
Control	2.34±0.60	2.34±0.30	2.34±0.80	2.34±0.20	2.11±0.20	2.11±0.60	2.11±0.70	2.11±0.20
Control 25	21.73±0.00	21.73±0.00	21.73±0.00	21.73±0.00	21.06±3.00	21.06±0.00	21.06±0.00	21.06±0.00
25 + EDTA	19.95±0.00	19.73±1.00	19.59±0.00	19.51±0.00	18.40±0.00	18.16±0.00	17.90±0.00	17.63±0.30
Control 50	44.33±0.20	44.33±0.20	44.33±0.20	44.33±0.20	43.50±6.00	43.50±6.00	43.50±6.00	43.50±6.00
50 + EDTA	35.41±0.08	35.27±3.00	35.12±6.00	34.94±0.00	32.33±0.00	31.20±0.00	30.90±0.00	30.46±5.00
Control 75	69.54±0.67	69.54±0.67	69.54±0.67	69.54±0.67	69.26±0.00	69.26±0.00	69.26±0.00	69.26±0.00
75 + EDTA	54.21±0.20	54.06±0.40	53.84±0.00	53.65±6.00	49.43±0.00	48.80±0.00	48.26±0.40	47.33±0.00
Control 100	81.36±0.70	81.36±0.70	81.36±0.70	81.36±0.70	74.96±7.00	74.96±7.00	74.96±7.00	74.96±7.00
100 + EDTA	78.56±0.02	78.29±0.40	78.12±0.00	77.86±0.00	74.35±0.00	74.20±0.00	74.02±6.00	72.94±5.00
Control 25	21.72±0.20	21.72±0.20	21.72±0.20	21.72±0.20	21.11±3.00	21.11±3.00	21.11±3.40	21.11±3.00
25 + C.A	22.09±0.20	21.95±0.60	21.81±7.00	21.67±0.00	21.90±0.00	21.76±0.00	21.61±0.00	21.47±0.30
Control 50	45.31±4.40	45.31±4.40	45.31±4.40	45.31±4.40	44.66±5.00	44.66±5.00	44.66±5.00	44.66±5.00
50 + C.A	42.87±0.40	41.87±0.60	41.72±0.40	41.57±7.00	38.66±0.00	38.40±0.00	38.13±0.00	37.86±0.00
Control 75	65.93±0.40	65.93±0.40	65.93±0.40	65.93±0.40	65.88±0.40	65.88±0.40	65.88±0.40	65.88±0.40
75 + C.A	63.49±0.40	63.33±0.30	63.15±0.00	62.81±6.00	58.91±0.20	58.77±0.00	58.62±0.00	58.48±0.00
Control 100	79.86±0.00	79.86±0.00	79.86±0.00	79.86±0.00	78.34±5.00	78.34±5.00	78.34±5.00	78.34±5.00
100 + C.A	79.41±0.40	79.26±0.00	79.12±6.00	78.97±7.00	74.02±0.00	73.87±0.00	73.73±0.00	73.58±0.00
Control 25	22.68±0.40	22.68±0.40	22.68±0.40	22.68±0.40	22.53±0.00	22.53±0.00	22.53±0.00	22.53±0.00
25 + EDTA and C.A	21.78±0.50	21.71±0.30	21.63±0.40	21.46±0.00	20.76±0.30	20.40±0.00	20.13±0.00	19.96±0.00

Control 50	42.16±0.30	42.16±0.30	42.16±0.30	42.16±0.30	39.43±6.00	39.43±6.00	39.43±6.00	39.43±6.00
50 + EDTA and C.A	40.41±0.10	40.27±7.00	40.11±0.30	39.97±0.00	36.49±0.30	36.26±0.00	36.08±0.30	35.90±5.00
Control 75	66.06±5.40	66.06±5.40	66.06±5.40	66.06±5.40	64.26±6.00	64.26±6.00	64.26±6.00	64.26±6.00
75 + EDTA and C.A	67.11±4.40	66.97±6.30	66.82±6.00	66.67±0.00	65.53±0.20	65.23±0.00	65.00±6.00	64.70±0.00
Control 100	86.73±0.50	86.73±0.50	86.73±0.50	86.73±0.50	81.33±6.00	81.33±6.00	81.33±6.00	81.33±6.00
100 + EDTA and C.A	84.31±0.50	84.17±5.00	84.02±5.00	83.87±0.00	81.80±0.40	81.13±0.60	80.80±0.30	80.60±0.78

Table 4.18.2: Remaining F concentration in the soil after harvesting the plantlets of *P. juliflora* for 90 and 120 days in different concentrations of F viz. control, 25, 50, 75, and 100 mg kg^{-1} NaF after treatment with chelates EDTA and CA

Fluoride content (mg kg^{-1})								
90 days					**120 days**			
Treatment	**5 mM**	**10 mM**	**15 mM**	**20 mM**	**5 mM**	**10 mM**	**15 mM**	**20 mM**
Control	2.00±0.40	2.00±0.60	2.00±0.30	2.00±0.72	1.34±0.56	1.34±0.80	1.34±0.89	1.34±0.36
Control 25	20.06±0.00	20.06±0.00	20.06±0.00	20.06±0.00	0.00±0.00	0.00±0.00	0.00±0.00	0.00±0.00
25+ EDTA	15.96±0.00	15.73±0.00	15.50±0.04	15.26±0.00	0.53±0.30	0.46±0.00	0.44±0.00	0.30±0.00
Control 50	41.86±0.00	41.86±0.00	41.86±0.00	41.86±0.00	15.37±0.30	15.37±0.30	15.37±0.30	15.37±0.30
50+ EDTA	25.96±0.50	25.76±0.60	25.56±0.50	25.36±0.10	12.46±0.00	12.15±0.45	12.09±0.30	12.02±0.47
Control 75	68.59±0.60	68.59±0.60	68.59±0.60	68.59±0.60	30.52±0.40	30.52±0.40	30.52±0.40	30.52±0.40
75+ EDTA	42.00±0.40	41.76±0.00	41.53±0.00	41.26±3.30	31.44±0.00	31.67±0.30	31.50±0.45	31.42±0.00
Control 100	67.90±0.60	67.90±0.60	67.90±0.60	67.90±0.60	47.67±0.36	47.67±0.36	47.67±0.36	47.67±0.36
100+ EDTA	60.56±0.40	60.36±0.00	60.16±0.00	59.63±0.70	34.10±0.40	33.86±0.30	33.06±0.00	31.62±0.00
Control 25	21.16±0.00	21.16±0.00	21.16±0.00	21.16±0.00	1.77±0.08	1.77±0.8	1.77±0.08	1.77±0.08
25+ C.A	21.15±0.40	21.07±0.00	20.94±0.00	20.81±6.00	0.43±0.40	0.20±0.00	0.00±0.00	0.00±0.00
Control 50	40.66±0.70	40.66±0.70	40.66±0.70	40.66±0.70	21.30±0.56	21.30±0.56	21.30±0.56	21.30±0.56
50+ C.A	28.36±0.40	27.76±0.00	27.53±4.00	27.30±0.80	25.51±0.40	25.34±0.30	25.19±0.78	25.03±0.00

Control 75	64.40±0.70	64.40±0.70	64.40±0.70	64.40±0.70	40.87±0.00	40.87±0.00	40.87±0.00	40.87±0.00
75 + C.A	42.33±0.20	42.03±0.00	41.80±0.00	41.53±0.00	28.67±0.40	28.28±0.83	27.88±0.89	27.12±0.00
Control 100	72.46±0.70	72.46±0.70	72.46±0.70	72.46±0.70	52.85±0.40	52.85±0.40	52.85±0.40	52.85±0.40
100 + C.A	58.50±0.50	58.26±0.00	58.03±0.00	57.83±0.00	44.49±0.00	44.31±0.00	44.07±0.20	43.65±0.00
Control 25	21.26±0.80	21.26±0.80	21.26±0.80	21.26±0.80	1.15±0.78	1.15±0.78	1.15±0.78	1.15±0.78
25 + EDTA and C.A	19.36±0.80	19.06±0.00	18.83±0.00	18.60±6.00	0.00±0.20	0.00±0.00	0.00±0.00	0.00±0.00
Control 50	36.40±0.20	36.40±0.20	36.40±0.20	36.40±0.20	24.30±0.10	24.30±0.10	24.30±0.10	24.30±0.10
50+ EDTA and C.A	32.50±4.00	32.23±0.00	32.03±0.02	31.66±0.30	10.93±0.40	3.82±0.00	3.78±0.00	3.74±0.00
Control 75	63.86±0.20	63.86±0.20	63.86±0.20	63.86±0.20	43.39±0.70	43.39±0.70	43.39±0.70	43.39±0.70
75 + EDTA and C.A	64.50±3.00	64.26±0.00	64.06±6.00	63.80±4.40	29.77±0.40	29.75±0.00	29.59±0.00	29.59±0.89
Control 100	73.03±0.20	73.03±0.20	73.03±0.20	73.03±0.20	64.06±0.20	64.06±0.20	64.06±0.20	64.06±0.20
100 + EDTA and C.A	75.43±2.00	75.20±0.20	74.96±0.87	74.70±0.10	46.63±0.00	46.55±0.00	46.52±0.00	46.15±0.00

Table 4.19.1: Translocation factor values (shoot/root concentration ratio) of *P. juliflora* in the soil for 30 and 60 days in different concentrations of F viz. control, 25, 50, 75 and 100 mg kg^{-1} NaF after treatment with chelates EDTA and CA

Fluoride content								
30 days					**60 days**			
Treatment	**5 mM**	**10 mM**	**15 mM**	**20 mM**	**5 mM**	**10 mM**	**15 mM**	**20 mM**
Control	0.80	0.80	0.80	0.80	3.69	3.69	3.69	3.69
Control 25	0.58	0.58	0.58	0.58	0.61	0.61	0.61	0.61
25 + EDTA	0.56	0.53	0.51	0.51	0.50	0.50	0.52	0.53
Control 50	0.75	0.75	0.75	0.75	0.96	0.96	0.96	0.96
50 + EDTA	0.30	0.30	0.30	0.30	0.30	0.29	0.30	0.30
Control 75	0.28	0.28	0.28	0.28	0.34	0.34	0.34	0.34

75+ EDTA	0.37	0.36	0.37	0.37	0.31	0.31	0.31	0.31
Control 100	0.53	0.53	0.53	0.53	0.45	0.45	0.45	0.45
100 + EDTA	0.17	0.18	0.18	0.19	0.18	0.17	0.17	0.17
Control 25	0.51	0.51	0.51	0.51	0.64	0.64	0.64	0.64
25+ C.A	0.34	0.32	0.30	0.29	0.43	0.40	0.39	0.37
Control 50	0.85	0.85	0.85	0.85	0.80	0.80	0.80	0.80
50+ C.A	0.29	0.28	0.28	0.28	0.23	0.24	0.25	0.26
Control 75	0.70	0.70	0.70	0.70	0.43	0.43	0.43	0.43
75 + C.A	0.25	0.25	0.25	0.27	0.32	0.31	0.31	0.31
Control 100	0.60	0.60	0.60	0.60	0.24	0.24	0.24	0.24
100+ C.A	0.35	0.35	0.35	0.35	0.35	0.35	0.35	0.35
Control 25	0.73	0.73	0.73	0.73	0.76	0.76	0.76	0.76
25 + EDTA and C.A	0.82	0.89	0.87	0.82	1.01	0.91	0.92	0.91
Control 50	0.70	0.70	0.70	0.70	0.96	0.96	0.96	0.96
50+ EDTA and C.A	0.55	0.54	0.53	0.52	0.47	0.48	0.48	0.49
Control 75	0.58	0.58	0.58	0.58	0.64	0.64	0.64	0.64
75 + EDTA and C.A	0.89	0.86	0.84	0.82	0.77	0.78	0.79	0.79
Control 100	0.40	0.40	0.40	0.40	0.63	0.63	0.63	0.63
100+ EDTA and C.A	0.40	0.40	0.39	0.39	0.40	0.39	0.40	0.40

Table 4.19.2: Translocation factor values (shoot/root concentration ratio) of *P. juliflora* in the soil for 90 and 120 days in different concentrations of F viz. control, 25, 50, 75 and 100 mg kg^{-1} NaF after treatment with chelates (EDTA and CA)

Fluoride content								
90 days					**120 days**			
Treatment	**5 mM**	**10 mM**	**15 mM**	**20 mM**	**5 mM**	**10 mM**	**15 mM**	**20 mM**
Control	0.86	0.86	0.86	0.86	0.83	0.83	0.83	0.83
Control 25	0.55	0.55	0.55	0.55	0.86	0.86	0.86	0.86
25+ EDTA	0.73	0.74	0.75	0.76	1.18	0.46	0.46	0.46
Control 50	0.82	0.82	0.82	0.82	0.88	0.88	0.88	0.88
50+ EDTA	0.48	0.48	0.49	0.50	1.03	1.02	1.01	1.00
Control 75	0.23	0.23	0.23	0.23	0.90	0.90	0.90	0.90
75 + EDTA	0.46	0.47	0.47	0.47	1.03	1.05	1.06	1.06
Control 100	0.43	0.43	0.43	0.43	0.98	0.98	0.98	0.98
100+ EDTA	0.56	0.56	0.56	0.56	1.23	1.23	1.25	1.28
Control 25	0.82	0.82	0.82	0.82	0.79	0.79	0.79	0.79
25+ C.A	0.51	0.51	0.49	0.47	0.85	0.85	0.85	0.86
Control 50	0.52	0.52	0.52	0.52	0.65	0.65	0.65	0.65
50+ C.A	0.94	0.97	0.97	0.97	1.00	0.99	0.99	0.98
Control 75	0.47	0.47	0.47	0.47	1.02	1.02	1.02	1.02
75 + C.A	0.76	0.76	0.76	0.77	0.97	0.98	0.99	1.01
Control 100	0.23	0.23	0.23	0.23	0.95	0.95	0.95	0.95
100 + C.A	0.70	0.70	0.71	0.71	1.02	1.02	1.02	1.02
Control 25	0.60	0.60	0.60	0.60	0.96	0.96	0.96	0.96
25 + EDTA and C.A	0.74	0.76	0.79	0.82	0.89	0.90	0.91	0.90
Control 50	0.85	0.85	0.85	0.85	0.93	0.93	0.93	0.93
50+ EDTA and C.A	0.41	0.42	0.43	0.43	0.70	0.98	0.98	0.98
Control 75	0.69	0.69	0.69	0.69	0.96	0.96	0.96	0.96
75 + EDTA and C.A	0.96	0.97	0.98	0.98	0.95	0.95	0.95	0.95
Control 100	0.55	0.55	0.55	0.55	0.94	0.94	0.94	0.94
100 + EDTA and C.A	0.33	0.34	0.34	0.35	0.96	0.96	0.96	0.97

Table 4.19.3: Bioaccumulation factor values (shoot/soil concentration ratio) of *P. juliflora* plants grown in soil under different concentrations of F viz. control, 25, 50, 75 and 100 mg kg^{-1} NaF soil for 30 and 60 days after treatment with chelates EDTA and CA

Fluoride content								
30 days					**60 days**			
Treatment	**5 mM**	**10 mM**	**15 mM**	**20 mM**	**5 mM**	**10 mM**	**15 mM**	**20 mM**
Control	0.68	0.68	0.68	0.68	0.72	0.72	0.72	0.72
Control 25	0.05	0.05	0.05	0.05	0.07	0.07	0.07	0.07
25+ EDTA	0.09	0.09	0.09	0.09	0.11	0.12	0.13	0.14
Control 50	0.05	0.05	0.05	0.05	0.07	0.07	0.07	0.07
50 + EDTA	0.09	0.09	0.09	0.09	0.12	0.13	0.14	0.14
Control 75	0.01	0.01	0.01	0.01	0.02	0.02	0.02	0.02
75+ EDTA	0.10	0.10	0.10	0.10	0.12	0.12	0.13	0.13
Control 100	0.07	0.07	0.07	0.07	0.10	0.10	0.10	0.10
100+ EDTA	0.04	0.04	0.04	0.04	0.05	0.05	0.05	0.05
Control 25	0.05	0.05	0.05	0.05	0.07	0.07	0.07	0.07
25 + C.A	0.03	0.03	0.03	0.03	0.00	0.04	0.04	0.04
Control 50	0.04	0.04	0.04	0.04	0.05	0.05	0.05	0.05
50 + C.A	0.04	0.04	0.04	0.04	0.05	0.05	0.06	0.06
Control 75	0.05	0.05	0.05	0.05	0.04	0.04	0.04	0.04
75+ C.A	0.03	0.03	0.03	0.04	0.06	0.06	0.06	0.06
Control 100	0.09	0.09	0.09	0.09	0.05	0.05	0.05	0.05
100 + C.A	0.06	0.06	0.06	0.06	0.09	0.09	0.09	0.09
Control 25	0.04	0.04	0.04	0.04	0.04	0.04	0.04	0.04
25+ EDTA and C.A	0.06	0.06	0.07	0.07	0.10	0.10	0.11	0.12
Control 50	0.07	0.07	0.07	0.07	0.13	0.13	0.13	0.13
50 + EDTA and C.A	0.08	0.08	0.08	0.08	0.11	0.12	0.12	0.00
Control 75	0.04	0.04	0.04	0.04	0.06	0.06	0.06	0.06
75+ EDTA and C.A	0.05	0.05	0.05	0.05	0.06	0.06	0.06	0.07
Control 100	0.04	0.04	0.04	0.04	0.08	0.08	0.08	0.08
100+ EDTA and C.A	0.05	0.05	0.05	0.05	0.06	0.06	0.06	0.06

Table 4.19.4: Bioaccumulation factor values (shoot/soil concentration ratio) of *P. juliflora* plants grown in soil under different concentrations of F viz. control, 25, 50, 75 and 100 mg kg^{-1} NaF soil for 90 and 120 days after treatment with chelates EDTA and CA

Fluoride content								
90 days					**120 days**			
Treatment	**5 mM**	**10 mM**	**15 mM**	**20 mM**	**5 mM**	**10 mM**	**15 mM**	**20 mM**
Control	0.95	0.95	0.95	0.95	1.99	1.99	1.99	1.99
Control 25	0.08	0.08	0.08	0.08	0.00	0.00	0.00	0.00
25+ EDTA	0.24	0.25	0.26	0.27	26.68	30.85	32.63	47.49
Control 50	0.08	0.08	0.08	0.08	1.57	1.57	1.57	1.57
50 + EDTA	0.30	0.30	0.31	0.32	2.90	2.98	3.01	3.04
Control 75	0.01	0.01	0.01	0.01	1.05	1.05	1.05	1.05
75 + EDTA	0.25	0.25	0.25	0.26	1.21	1.20	1.21	1.22
Control 100	0.14	0.14	0.14	0.14	0.86	0.86	0.86	0.86
100+ EDTA	0.23	0.23	0.23	0.24	1.32	1.33	1.37	2.07
Control 25	0.08	0.08	0.08	0.08	6.91	6.91	6.91	6.91
25 + C.A	0.06	0.06	0.06	0.06	49.61	0.10	0.00	0.77
Control 50	0.07	0.07	0.07	0.07	1.15	1.15	1.15	1.15
50 + C.A	0.37	0.39	0.40	0.41	1.30	1.71	1.33	1.74
Control 75	0.05	0.05	0.05	0.05	0.82	0.82	0.82	0.82
75 + C.A	0.33	0.34	0.34	0.35	1.57	1.60	1.62	1.67
Control 100	0.07	0.07	0.07	0.07	0.89	0.89	0.89	0.89
100 + C.A	0.29	0.29	0.30	0.30	1.76	1.76	1.77	1.80
Control 25	0.00	0.00	0.00	0.00	1.49	1.49	1.49	1.49
25+ EDTA and C.A	0.12	0.13	0.14	0.15	0.00	0.00	0.00	0.00
Control 50	0.17	0.17	0.17	0.17	0.21	0.21	0.21	0.21
50 + EDTA and C.A	0.15	0.16	0.16	0.17	3.30	9.50	9.63	9.77
Control 75	0.07	0.07	0.07	0.07	0.12	0.12	0.12	0.12
75+ EDTA and C.A	0.08	0.08	0.08	0.08	2.22	2.23	2.25	2.25
Control 100	0.13	0.13	0.13	0.13	0.17	0.17	0.17	0.17
100 + EDTA and C.A	0.08	0.08	0.08	0.08	1.15	1.15	1.16	1.17

The application of chelates with F concentrations led to a significant increase when 20 mM EDTA and CA were applied. Chelates EDTA, CA and combined EDTA+CA increased uptake of the F due to increased permeation through cell wall pores and transport proteins in membranes. Also, chelates have the ability to bind to carboxylic groups in the apoplasm as increased uptake of F (Geebelen et al., 2002).

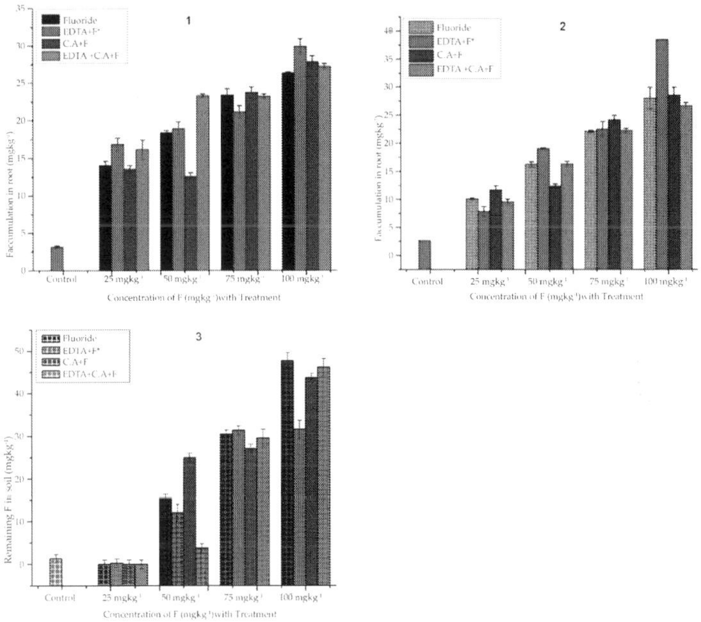

Figure 4.18: Effect of chelates (EDTA and CA) on root and shoot uptake under F stress for 120 days (1-root, 2- shoot uptake and 3-remaining F in soil, EDTA-Ethylene diamine tetraacetic acid and CA-Citric acid where significant value*).

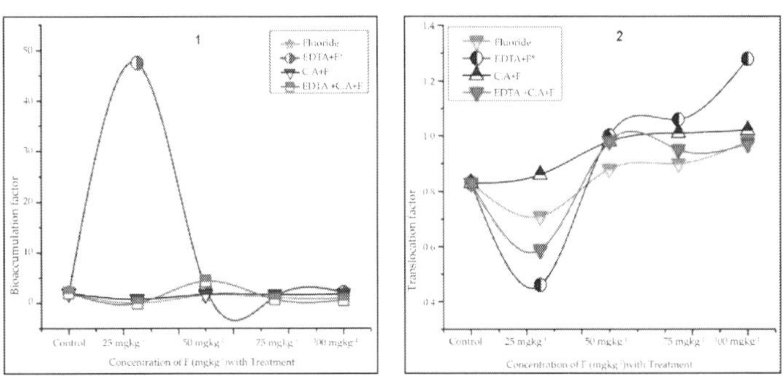

Figure 4.19: Effect of chelates (EDTA and CA) on bioaccumulation and translocation factor under F stress for 120 days (1-Bioaccumulation factor and 2-Translocation factor, EDTA-Ethylene diamine tetraacetic acid and CA-Citric acid where significant value*).

4.15 MINERAL CONTENT IN PLANT PARTS (ROOT, SHOOT AND LEAVES) TREATED WITH THE MICROBIAL CONSORTIUM (P.F AND P.A) UNDER DIFFERENT CONCENTRATION OF F

The analysis of N, P and K in root, shoot and leaves of *P. juliflora* is shown in Figure 4.20. The N, P and K amountsignificantly increased (p=0.98) when the microbial consortium level increased is observed in Figure 4.20. The N, P and K concentration in root under P.F treatment was found to be (7.96 kg ha^{-1}, 3.50 mg kg^{-1} and 21.06 kg ha^{-1}) higher in comparison to the treatment with P.A of*P. juliflora*. However, the N, P and K levels were affected by different F treatments (25, 50, 75 and 100 mg kg^{-1}), but was enhanced when the microbes treatment was given to the plants. Similar observations were obtained in white lupine with Cd treatment; significantly decreases in shoot K contents of white lupine (*Lupinus albus*) with 18 and 45 μM Cd treatments were also reported by Sharma and Pant (1994). However, Fe, Mn, Zn and Cu concentrations in the root significantly increased (p=1.00) with microbial treatment. The mean values of micronutrients (Fe, Mn, Zn and Cu) were found as 76.06, 27.80, 86.14 and 29.78mg kg-1 respectively.

The N, P and K were found as 9.66 kg ha^{-1}, 2.58 mg kg^{-1} and 20.06kgha^{-1} in the shoot respectively and it is observed that enhanced with microbe treatments by P.F. The available micronutrients also enhanced in the case of Fe, Mn, Zn and Cu (95.06, 16.56, 72.14 and 40.11mgkg^{-1}) respectively. Meanwhile, in leaves N, P and K were found at lower side viz (16.56 kg ha^{-1}, 6.56 mg $kg^{-1}$and 13.95 kg ha^{-1}) in comparison to the root and shoot.

Previous studies suggested that different kinds of nutrients were decreased including calcium, zinc, magnesium, phosphorus, potassium, iron and cooper under heavy metal stress (Ribeiro *et al.*, 2013). Heavy metals such as Cu, Zn, Al, Cd, Ni, Hg and As have negative effects on macronutrient and micronutrients (Mariano and Keltjens, 2005).

Heavy metals bind to all calcium binding sites on the cell surface at low pH 4.5 (Hossain *et al.*, 2014). High amount of heavy metals were exposed to the inhibition of calcium. It inhibits the influx of calcium ion at 100 mM aluminum (Al) (Nichol *et al.*, 1993).

Aluminum ions interfere with the action of guanosine 5' triphosphate (GTP) binding protein as well as inhibit calcium ion uptake by binding the verapamil-specific channel (Rengel and Elliott, 1992). The uptake of calcium ions was decreased in beech plants due to the combination of nitrogen and aluminum at high concentrations (Bengdtsson *et al.*, 1994). Potassium channel influx is inhibited due to the toxicity of heavy metals like Aluminium. Active pathway involvement of potassium uptake is also inhibited by a high concentration of aluminum (Hossain *et al.*, 2014). Durum wheat is more tolerant against aluminum toxicity but potassium ion concentration was decreased in wheat under aluminum concentration (Zsoldos *et al.*, 2000).

Decreased concentration of magnesium ions was found in roots and shoots by increasing the concentration of heavy metals (Huang and Bachelard, 1993). Iron concentration decreased at a pH of 4.5 when they were treated with different concentrations of heavy metals and also significantly affected root growth (Moustakas *et al.*, 2008).

Therefore, there are chances that calcium channel and active pathway interferes with the action of F with P.F and P.A. to increase mineral content with a specific channel like GTP and Verapamil.

Therefore, P.F has a positive effect on mineral content as shown in (Figure 4.20). It seems that F has combative interactions especially at high concentrations with mineral content in the plant, which led to decreased mineral level.

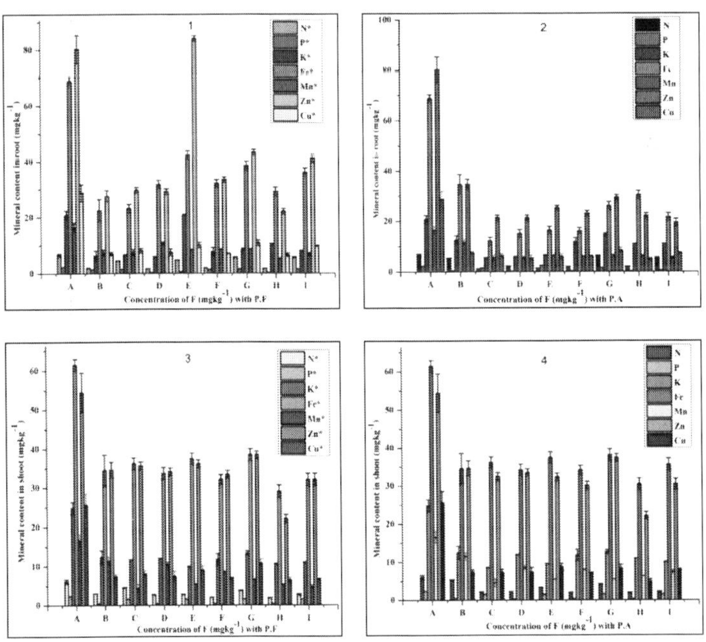

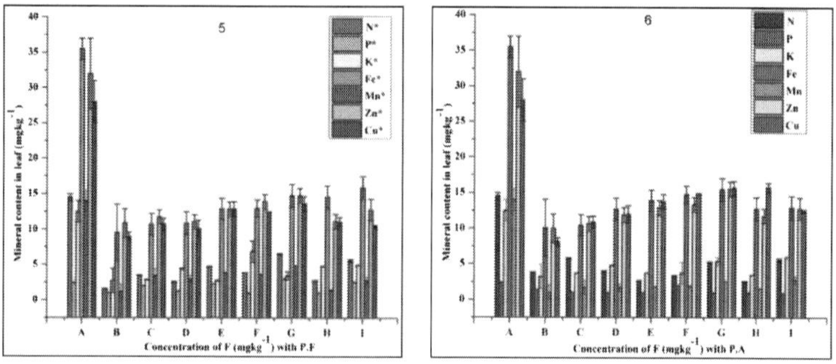

Figure 4.20: Effect of microbial consortium (P.F and P.A) on mineral contents of *P. juliflora* seedlings under different concentration of F viz., control, 25, 50, 75 and 100 NaF mg kg^{-1}root (1-2), shoot (3-4), leaf (5-6) for 120 days - A-control, B-25 mg kg^{-1}, C-25 mg kg^{-1}+ P.F, D-50 mg kg^{-1}, E-50 mg kg^{-1}+P.F, F-75 mg kg^{-1}, G-75 mg kg^{-1}+P.F, H-100 mg kg^{-1}, I-100 mg kg^{-1}+P.F, same set of experiments conducted with PA (P.F- *Pseudomonas fluorescens*, P.A- *Pseudomonas aeruginosa* significantvalue*).

4.16 MINERAL CONTENT IN PLANT PARTS (ROOT, SHOOT AND LEAVES)TREATED WITH CHELATES (EDTA AND CA) UNDER DIFFERENT CONCENTRATION OF F

The toxic effects of F on mineral contents by increasing concentration in plants are shown in (Figure 4.21). The EDTA, CA and combined EDTA with C.A have significantly increased the accumulation of plant nutrients at20 m Mkg^{-1}. The content of these nutrients i.e N, P, and K were measured as shown in (Figure 4.21).

The N, P and K concentration had increased significantly. The N concentration in root was (4.55 kg ha^{-1}, 4.44 mg kg^{-1} and 2.67kgha^{-1}) under different treatments of EDTA, CA and combined EDTA with C.A. respectively.

However, N concentration without chelate treatment was affected by F concentration and ranges from4.07 to 1.65 kg ha^{-1}.

The P concentrations in root were as follows (4.65, 2.49 and 1.44mgkg^{-1}) whereas without treatment the P level ranges from (3.00, 2.35 and 1.35mg kg^{-1}) respectively. The K concentration in root significantly increased with the increased concentration of chelating agent (Figure 4.21). Decreasing

K uptakes of wheat seedlings with Cd treatments were also reported by Veselov *et al.*(2003). Such decreases in K concentrations might be related to ATPase, responsible for active K uptake (Lindberg and Wingstrand, 1985). The association of *P. juliflora* plants with EDTA significantly maintained the uptake of nutrients by roots and subsequently translocate it into shoot and leaves.

Hence, from the present study, it is concluded that EDTA has adverse the effects of F contamination on the physiological status of plants by improving the availability and absorption of the nutrients for plants.

The Fe concentration in root had significantly increased with EDTA. The Fe is a crucial plant nutrient which maintains plant growth and plays an important role in metabolic activities.

The mean value of micronutrients Fe, Mn, Zn, and Cu were shown in (Figure 4.21).

The study showed that the *P. juliflora* plants grown in F treated soils showed a significant decrease in the uptake of some essential macronutrients and micronutrients.

The concentration of Cu decreased significantly as compared to control with increasing F concentrations.

The available N, P and K in the shoot were also affected which is shown in Figurc 4.21.

The concentrations of micronutrients (Fe, Mn, Zn and Cu)were significantly increased with the given treatment by EDTA. Meanwhile, the concentration of macronutrients and micronutrients in leaves were significantly increased at 20 mM kg^{-1} EDTA.

Mineral content viz Fe, Mn and Cu were increasing with EDTA application.

Similar observations were found in the case of contaminated field sites, excessive heavy metals are often associated with a shortage of available mineral nutrients(Shetty *et al.*, 1994a,b).

Earlier results have indicated that EDTA is a powerful mobilizing agent for increasing Pb concentrations in the soil solution and in plant shoots (Huang *et al.*, 1997; Chen *et al.*, 2004).

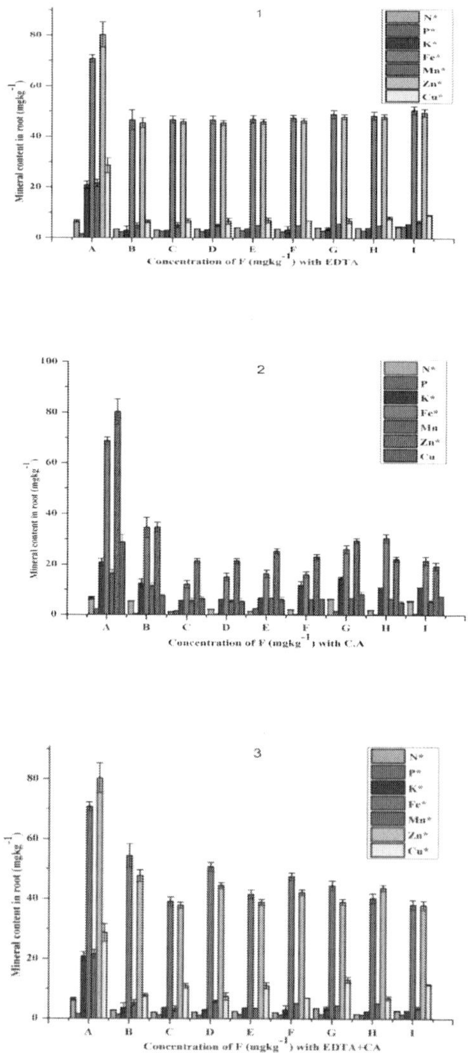

Figure 4.21.A: Effect of chelates (EDTA, CA and combined EDTA+CA) on mineral contents of *P. Juliflora* seedlings under different concentration of F viz., control, 25, 50, 75 and 100 NaF mg kg^{-1}) root (1-2-3) for 120 days. A-control, B-25 mg kg^{-1}NaF, C-25 mg kg^{-1}NaF + EDTA, D-50 mg kg^{-1}NaF, E-50 mg kg^{-1}NaF +EDTA, F-75 mg kg^{-1}NaF, G-75 mg kg^{-1}NaF+EDTA, H-100 mg kg^{-1}NaF, I-100 mg kg^{-1}NaF+EDTA,same set of experiments conducted withCA and EDTA+CA (EDTA-Ethylene diamine tetraacetic acid and CA-Citric acid, where significant value*).

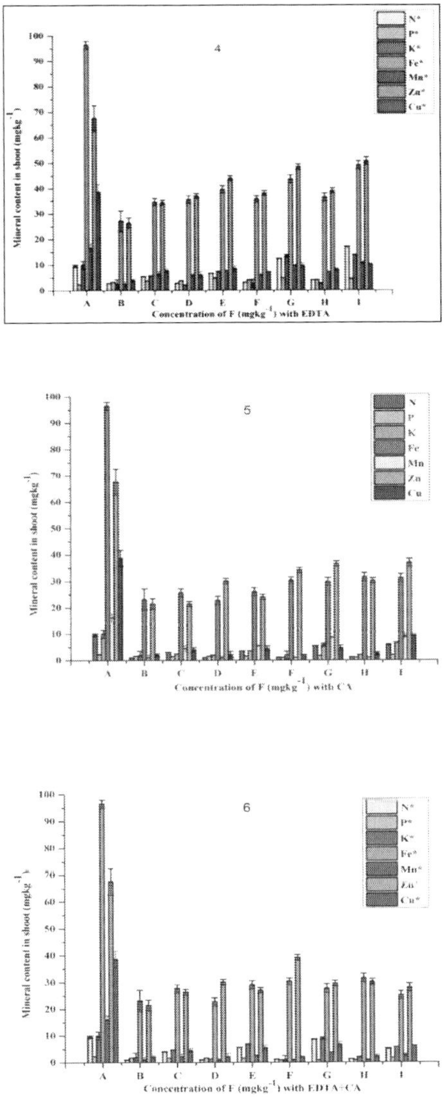

Figure 4.21.B: Effect of chelates (EDTA, CA and combined EDTA+CA) on mineral contents of *P. Juliflora* seedlings under different concentration of F viz., control, 25, 50, 75 and 100 NaF mg kg^{-1}) shoot (4-5-6) for 120 days. A-control, B-25 mg kg^{-1}NaF, C-25 mg kg^{-1}NaF + EDTA, D-50 mg kg^{-1}NaF, E-50 mg kg^{-1}NaF +EDTA, F-75 mg kg^{-1}NaF, G-75 mg kg^{-1}NaF+EDTA, H-100 mg kg^{-1}NaF, I-100 mg kg^{-1}NaF+EDTA, same set of experiments conducted with CA and EDTA+CA (EDTA-Ethylene diamine tetraacetic acid and CA-Citric acid, where significant value*).

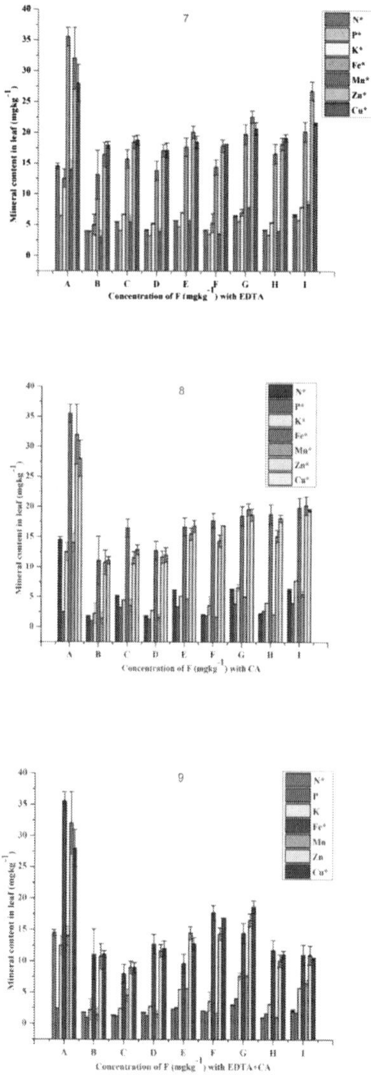

Figure 4.21.C: Effect of chelates (EDTA, CA and combined EDTA+CA) on mineral contents of *P. Juliflora* seedlings under different concentration of F viz., control, 25, 50, 75 and 100 NaF mg kg^{-1}) leaf (7-8-9) for 120 days. A-control, B-25 mg kg^{-1}NaF, C-25 mg kg^{-1}NaF + EDTA, D-50 mg kg^{-1}NaF, E-50 mg kg^{-1}NaF +EDTA, F-75 mg kg^{-1}NaF, G-75 mg kg^{-1}NaF+EDTA, H-100 mg kg^{-1}NaF, I-100 mg kg^{-1}NaF+EDTA, same set of experiments conducted with CA and EDTA+CA (EDTA-Ethylene diamine tetraacetic acid and CA-Citric acid, where significant value*).

4.17 MINERALCONTENT IN SOIL AFTER MICROBES (P.F AND P.A) AND CHELATES(EDTA AND CA) TREATMENT UNDER DIFFERENT CONCENTRATION OF F

The physical and chemical properties of the soil are shown in Tables 4.20.1 and 2. The pH was found to be 6.37. The pH of soil samples was almost alkaline. The soil pH was within the suitable range (5.0-7.0), ensures the high bioavailability of most essential nutrients required for plant growth and development (Violante *et al*.,2010). In addition, soil pH plays a vital role in the absorption and retention of F. The uptake of F is greatly affected if the soil pH does not fall within the optimum range. The electrical conductivity is a numerical expression of ability of an aqueous solution to carry electrical current. The E.C was found to be minimum 1.98 at 25 mg kg-1 NaF and maximum 4.25 dSm^{-1} with P.F and 20 mM Kg^{-1} EDTA with 100 mg kg-1 NaF.Furthermore, plants need macronutrients (N, P and K) in huge amounts to sustain their healthy growth and development. The total N content of the soil sample was obtained in the range of 140 to128.64 $kgha^{-1}$. Nitrogen promotes rapid growth of plants by helping in wider leaves and long stems development (Baloch *et al*.,2014).It is an essential mineral for protein and chlorophyll structure. The phosphorus (P) concentrations in the soil were found in the range of 28.45 and 29.43mg kg^{-1}. It plays an important role in promoting and initiating root growth at the early stage of the plant development for optimum absorption of soil nutrients and water. In addition, P assists in efficient water intake of plants through roots, contributes to genetic material formation and participates in plant reproduction and cell division (Baloch *et al*.,2014). Moreover, potassium (K) was present in the soil with concentration varied from 208 and 150 kg ha^{-1}. Thus, it is necessary for the development of thick and strong cell walls to provide structural support to the plants. It also regulates the opening and closing of stomata at the leaf surface to allow gaseous exchange such as oxygen, carbon dioxide and water vapor in and out of the leaf structure. Therefore, K is very important for plant survival as photosynthesis and transpiration are critical biochemical processes involving gaseous exchange (Khan *et al*.,2014). Micronutrient iron (Fe) was found from 6.73 to 6.55$mgkg^{-1}$. The plant growth and development also depend on Fe which promotes the chlorophyll formation. It is a part of many enzymes necessary for the formation of starch and proteins. Mn accumulation in the root, shoot and leaves was more than (5.83 and 5.89$mgkg^{-1}$). While, Zinc accumulation in the aboveground biomass was more than 4.91$mgkg^{-1}$.Copper accumulation in the aboveground biomass was found low 5.83 (Table 4.20.1 and 4.20.2)

(Baloch *et al.*,2014).

Table 4.20.1: Effect of microbes P.F and P.A on soil nutrients of *P. juliflora* seedlings were grown in soil for 120 days in different concentrations of F viz. control, 25, 50, 75 and 100 mg kg^{-1} NaF in a pot experiment

T	pH		E.C		N		P		K		Fe		Mn		Zn		Cu	
T0	7.68	±0.08	2.60	±0.90	116.70	±2.78	15.43	±1.07	114.00	±1.00	4.52	±1.89	3.67	±0.00	3.68	±0.90	2.55	±0.19
T1	7.73	±0.09	1.98	±1.00	118.74	±2.56	12.82	±0.06	152.00	±1.78	3.31	±0.00	2.77	±0.00	4.08	±0.09	1.11	±0.17
T2	6.24	±0.17	2.88	±0.78	114.75	±2.45	25.83	±0.05	163.00	±1.37	5.53	±0.00	4.37	±0.90	4.24	±0.00	5.56	±0.44
T3	7.72	±0.06	2.19	±0.37	119.54	±1.90	16.35	±2.08	158.00	±1.90	3.54	±1.00	2.79	±0.00	4.68	±0.10	2.09	±0.78
T4	5.84	±0.07	3.23	±0.90	124.36	±1.67	25.93	±0.06	173.00	±1.00	5.60	±0.78	4.14	±0.78	4.91	±0.20	5.76	±0.23
T5	7.82	±0.30	2.57	±0.20	120.75	±1.56	16.45	±2.69	160.00	±1.00	3.75	±0.37	2.88	±0.07	4.75	±0.08	2.18	±0.90
T6	5.72	±0.67	3.63	±0.40	125.98	±1.09	25.85	±2.00	175.00	±1.00	5.72	±0.08	4.53	±0.90	4.83	±0.07	5.80	±0.22
T7	7.73	±0.39	2.98	±0.20	121.50	±1.70	17.78	±3.90	168.00	±1.78	3.82	±0.05	2.95	±0.50	4.85	±0.10	2.57	±0.09
T8	5.82	±0.29	3.45	±0.78	140.64	±2.09	28.45	±1.00	208.00	±1.37	6.73	±0.08	5.83	±0.08	4.91	±0.20	5.83	±0.07
T9	7.83	±0.22	2.84	±0.37	115.60	±2.78	16.76	±2.78	106.00	±1.90	3.55	±1.09	2.68	±0.78	2.74	±0.68	1.76	±0.05
T10	5.72	±0.28	2.63	±0.90	122.90	±2.09	26.75	±0.37	125.00	±1.00	3.45	±0.00	5.23	±0.00	4.23	±0.08	4.18	±0.80
T11	6.45	±0.00	2.96	±0.10	116.50	±2.78	18.54	±4.90	109.00	±2.88	3.55	±0.90	2.78	±0.00	2.84	±0.19	1.87	±0.19
T12	5.97	±0.07	2.84	±0.90	123.90	±2.56	26.67	±1.00	145.00	±3.00	4.54	±1.00	5.23	±0.00	4.33	±0.00	4.15	±0.17
T13	7.67	±1.08	3.04	±0.00	118.44	±2.45	19.23	±1.00	110.00	±1.00	5.03	±1.09	2.57	±0.17	2.95	±0.10	2.86	±0.44
T14	5.46	±0.09	3.44	±0.78	125.66	±1.90	26.76	±1.00	167.00	±1.78	5.65	±0.10	6.35	±0.18	4.65	±0.09	4.16	±0.78
T15	7.77	±0.17	3.20	±0.37	119.57	±1.67	20.87	±0.78	115.00	±1.37	5.34	±0.90	2.39	±0.37	3.61	±0.00	2.96	±0.23
T16	5.95	±1.06	3.34	±0.90	126.90	±1.56	29.46	±0.37	200.00	±1.90	6.76	±1.00	6.65	±0.90	4.75	±0.10	4.60	±0.90

Note: T: Treatment, T0: Control, T1: Control 25, T2: 25 + P.F, T3: Control 50, T4: 50 + P.F, T5: Control 75, T6: 75 + P.F, T7: Control 100, T8: 100 + P.F, T9: Control 25, T10: 25 + P.A, T11: Control 50 , T12: 50+ P.A, T13: Control

75, T14: 75+ P.A, T15: Control 100, T16: 100 + P.A, P.F: Pseudomonas fluorescens; P.A: Pseudomonas aeruginosa. Individual readings n = 3 for each plant were averaged and are presented with ±SD. Significant correlation was found between nutrients at 0.05 level of significance. E.C: Electrical conductivity; N: Nitrogen; P: Phosphorus; K: Potassium; Fe: Iron; Mn: Maganase; Zn: Zinc; Cu: Copper mg kg^{-1}.

Table 4.20.2: Effect of chelates EDTA and CA on soil nutrients of *P. juliflora* seedlings were grown in soil for 120 days in different concentrations of F viz. control, 25, 50, 75 and 100 mg kg^{-1} NaF in a pot experiment

T	pH		E.C		N		P		K		Fe		Mn		Zn		Cu	
T0	7.68	±0.78	2.60	±0.78	116.70	±1.78	15.43	±0.67	114.00	±1.78	4.52	±0.90	3.67	±0.08	3.68	±0.02	3.55	±0.01
T1	6.34	±2.37	2.48	±0.90	110.75	±1.67	16.81	±0.45	116.00	±1.67	2.81	±0.00	2.83	±0.06	2.78	±0.03	1.99	±0.18
T2	6.62	±3.90	3.48	±0.78	112.76	±5.78	18.85	±0.24	134.00	±5.78	4.05	±0.28	4.35	±0.00	4.44	±0.01	4.02	±0.19
T3	6.31	±1.00	2.59	±0.06	109.54	±4.67	12.33	±0.90	108.00	±4.67	2.61	±0.10	2.73	±0.90	3.73	±0.78	1.89	±0.00
T4	6.57	±2.90	3.53	±0.09	124.33	±3.67	20.92	±0.00	135.00	±3.67	4.20	±0.90	5.12	±0.00	4.91	±0.89	4.12	±0.10
T5	6.35	±2.67	2.57	±0.90	107.72	±6.90	12.16	+±0.68	106.00	±6.90	1.73	±0.00	2.43	±0.78	3.72	±0.00	1.58	±0.00
T6	6.42	±2.78	3.83	±0.00	125.96	±5.00	25.76	±0.90	143.00	±5.00	4.73	±0.78	5.23	±0.37	4.83	±0.90	4.14	±0.18
T7	6.33	±2.80	2.58	±0.90	106.56	±4.67	11.94	±0.00	104.00	±4.67	1.66	±0.37	2.36	±0.90	3.67	±0.00	1.57	±0.17
T8	6.37	±0.67	4.25	±1.00	128.64	±3.45	29.43	±0.08	150.00	±3.45	6.55	±0.90	5.89	±0.00	4.91	±0.78	5.73	±0.00
T9	6.21	±0.89	2.64	±0.78	105.67	±2.18	16.72	±0.00	116.00	±0.18	1.55	±0.20	2.98	±0.68	3.76	±0.37	2.68	±0.08
T10	5.68	±0.78	3.43	±0.90	112.92	±3.78	16.86	±0.78	128.00	±0.78	3.09	±0.40	3.23	±0.78	4.23	±0.00	2.18	±0.05
T11	6.36	±0.80	2.66	±0.08	104.57	±2.37	16 52	±0.90	111.00	±0.37	1.45	±0.20	2.87	±0.89	3.67	±0.08	2.66	±0.18
T12	6.35	±0.89	3.44	±0.05	113.94	±2.90	20 64	±0.78	129.00	±0 90	4.27	±0.78	3.33	±0.08	4.38	±0.05	3.15	±0.19
T13	6.12	±0.56	2.54	±0.00	103.47	±1.00	15.82	±0.00	110.00	±1.00	1.43	±0.37	2.99	±0.78	3.65	±0.78	2.55	±0.00
T14	6.35	±0.09	3.34	±0.10	115.63	±3.90	21.75	±0.78	133.00	±0.90	5.12	±0.90	3.35	±0.56	4.68	±0.89	3.16	±0.90
T15	6.28	±0.70	2.73	±0.08	102.55	±1.00	12.85	±0.37	108.00	±1.00	1.34	±0.10	2.70	±0.45	3.61	±0.00	2.54	±0.10

T16	6.47	±1.09	3.24	±0.08	116.92	±4.78	24.42	±0.67	144.00	±0.78	5.29	±0.88	4.10	±0.25	4.77	±0.90	3.54	±0.18
T17	6.21	±0.18	2.64	±0.56	105.62	±2.67	16.74	±0.15	116.00	±0.67	1.55	±0.50	2.98	±0.05	3.76	±0.00	2.68	±0.89
T18	4.68	±0.09	3.13	±0.20	102.91	±1.78	18.86	±0.14	118.00	±0.18	2.09	±0.65	2.23	±0.08	3.23	±0.78	2.18	±0.00
T19	6.36	±0.94	2.66	±0.18	104.55	±1.67	15.57	±0.78	111.00	±2.78	1.45	±0.68	2.87	±0.07	3.67	±0.89	2.66	±0.10
T20	6.35	±0.90	3.04	±0.07	103.96	±5.78	16.62	±0.37	119.00	±2.00	2.27	±0.87	2.21	±0.08	3.38	±0.00	2.15	±0.01
T21	6.12	±0.56	2.54	±0.03	103.47	±4.67	14.85	±0.67	106.00	±2.67	1.43	±0.56	2.99	±0.06	3.65	±0.90	2.55	±0.18
T22	5.35	±1.78	3.14	±0.05	105.67	±3.67	15.72	±0.45	123.00	±2.56	3.12	±0.90	2.15	±0.00	3.88	±0.00	2.10	±0.19
T23	6.28	±0.37	2.73	±0.06	102.52	±6.90	12.82	±0.67	105.00	±3.67	1.34	±1.00	2.00	±0.90	2.00	±0.02	2.54	±0.10
T24	6.47	±1.90	3.04	±0.08	106.91	±5.00	14.45	±0.45	124.00	±1.78	3.20	±0.28	2.10	±1.00	2.77	±0.03	2.60	±0.10

Note: T: Treatment, T0: Control, T1: Control 25, T2: 25 + EDTA, T3: Control 50 , T4: 50 + EDTA, T5: Control 75 , T6: 75+ EDTA, T7: Control 100 , T8: 100+ EDTA, T9: Control 25, T10: 25 + CA, T11: Control 50, T12: 50+ CA, T13: Control 75 , T14: 75 + CA, T15: Control 100, T16: 100 + CA, T17: Control 25, T18: 25 + EDTA+CA, T:19 Control 50, T20: 50 + EDTA+CA, T21: Control 75 , T22: 75 + EDTA+CA, T23: Control 100 , T24: 100 + EDTA+CA. EDTA: Ethylene diamine tetraacetic acid; C.A: Citric acid. Individual readings n = 3 for each plant were averaged and are presented with ±SD. Significant correlation was found between nutrients at 0.05 level of significance. E.C: Electrical conductivity; N: Nitrogen; P: Phosphorus; K: Potassium; Fe: Iron; Mn: Maganase; Zn: Zinc; Cu: Copper mg kg^{-1}.

4.18 GERMINATION PERCENTAGE AND GROWTH PARAMETERS AFTER TREATMENT WITH *PSEUDOMONAS FLUORESCENS* AND ETHYLENE DIAMINE TETRAACETIC ACID IN THE FIELD EXPERIMENT

The germination percentage was 45 percent in P.F and 55 percent in EDTA at 100mg kg^{-1}NaF. The effect of P.F and EDTA on plant growth was observed (Figure 4.22). The plant biomass was obtained as shown in(Figure s 4.22 and 4.30). Although the plant biomass was significantly ($p \leq 0.05$) affected at 100 mg kg^{-1}NaF. A decrease in root and shoot length of *P. juliflora* plants at 100 mg kg^{-1} NaF treatment was observed. Both shoot length and its biomass were enhanced after givenP.F and EDTA treatment. Root length and its biomass were also increased after P.F and EDTA treatment. The root length, shoot length and biomass of plant after treatment are shown in Figure 4.22.Similar results of an increase in biomass were reported in previous literature(Badr *et al.*, 2017). An increase in plant biomass along with root and shoot length may be due to the enhancement in micronutrients and macronutrients in the plant after *P. fluorescens* and Ethylene diamine tetraacetic acid treatment. It was reported that macronutrients are responsible for the cell division, cell signaling and growth, while the micronutrients help in the plant metabolism (Ribeiro *et al.*, 2013; Radoslow and Ciecko, 2017). Plant growth- promoting rhizosphere (PGPR) has more than one mechanism to increase plant growth and production of such metabolites enzyme, bioactive factors as well as growth promoters. F-tolerant bacteria having plant growth -promoting traits identified from the F contaminated soil was found to enhance plant biomass and increase the mobility of fluoride present in soil. Our research showed a positive effect of P.F and EDTAon micro and macronutrient levels. The plant growth can be persistent only if the micro and macronutrients in the soil were under the average concentration levels. Therefore, plants need a balanced nutrient supply to enhance their growth as well as for maintaining the biological and chemical reactions (Xia *et al.*, 2017).

4.19 CHLOROPHYLL CONTENTS AFTER TREATMENT WITH *PSEUDOMONAS FLUORESCENS* AND ETHYLENE DIAMINE TETRAACETIC ACID IN THE FIELD EXPERIMENT

The chlorophyll content of *P. juliflora* leaves was reduced as shown in

(Figure 4.23 and 4.29) under F stress as compared to control and further increased after treatment with P.F and EDTA such as Chl a (1.072mgg^{-1} fw), Chl b (0.512mgg^{-1} fw), total chl a+b (1.584mgg^{-1} fw), Chl a/b (2.093mgg^{-1} fw) and biostimulation of EDTA was Chl a (0.989mgg^{-1} fw), Chl b from (0.534mgg^{-1} fw), total Chl a+b (1.523mgg^{-1} fw), Chl a/b from (1.852mgg^{-1} fw)at 100 mg kg-1 of NaF.Fluoride accumulation is gradually increased with time and inhibits the photosynthesis process in plants when we applied more than this concentration at 100 mg kg^{-1} F concentration. In this process, F will move through transpiration from roots or through stomata and accumulate in margins of the leaf. Fluoride is considered phytotoxic and negatively affects respiration, germination, growth, amino acids, proteins, photosynthesis, stromal enzymes and reproduction (Garrec *et al.*, 1981; Elloumi *et al.*, 2005). It often inhibits a number of enzymes that are required for adsorption of ions such as Mn^{2+}, Ca^{2+}, Mg^{2+} (Panda, 2015). The sensitivity and toxicity to the plants were found after screened F concentrations exceed more than 100 mg kg^{-1} F. Similar results were found that reduction in chlorophyll content after certain period plant tissue start to die due to high accumulation of F (Landis *et al.,* 2011; Kumar *et al.,* 2013).

4.20 ANTIOXIDANTENZYME ACTIVITIES AFTER TREATMENT WITH *PSEUDOMONAS FLUORESCENS* AND ETHYLENE DIAMINE TETRAACETIC ACID IN A FIELD EXPERIMENT

Antioxidant activity of plant *P.juliflora* was enhanced after the treatment of F. Hydrogen peroxide behaved both as an oxidizing and reducing agent in plant organs. Peroxidase (POD), catalase (CAT) and superoxidase dismutase (SOD) activity was calculated as 0.135, 0.158, 0.155 in P.F, 0.158, 0.156, and 0.170 µgg^{-1}fw in case of EDTA at 100 mg kg-1 NaF treatment. The hyperactivity of POD under F stress indicated the scavenging activity of H_2O_2 generated throughout the action of photorespiration in plant cells. CAT activity scavenges H_2O_2 by breaking it directly into water and oxygen. The increase in antioxidant activity was observed in plants under F stress and reduced with given treatment of P.F and EDTA. The same results were found in Mycorrhizal fungi effect on plant antioxidant enzymes by the addition of earthworm under cadmium concentration in soil. Similar results were also observed in *Triticum aestivum* under saline stress (Heidari, 2009; Gill *et al.*, 2011).

4.21 MICROBE AND CHELATE ASSISTED PHYTOREMEDIATION IN THE FIELD EXPERIMENT

Pseudomonas fluorescens (P.F) and Ethylene diamine tetraacetic acid (EDTA) weresignificantly increased bioaccumulation and translocation factors in field- scale treatment (field experiment design is shown in Figure 4.25). The effects of P.F on F removal from soil were shown (Figure 4.25 and 4.31). Based on the pot experiment results, we can state that the efficacy of P.F and EDTA was higher approx (50%) as compared to field experiment results. The lower effects of microbial consortium in the field experiment as in comparison to laboratory scale may be due to a low amount of microbial consortia for solubility, insufficient oxygen supply and low solubility of F ions to form a complex with other metals in soil. *P. juliflora* had accumulated by P.F 33.14, by EDTA 31.45in root and 34.56 in P.F, 35.67 mg $kg^{-1}$of NaFshoot dry tissue in EDTA, respectively. Translocation factor (TF) and bioaccumulation factor (BF) obtained were 1.04 and 1.06 which justify the high F accumulation in shoot tissues than in root tissues (Figure 4.26 and 4.32).

The present study calculated the BF as the ratio between the total F content in the shoot against total F in the soil. In this study, BF obtained was more than one which reflects the potential of *P. juliflora* for phytoremediation of F. As for efficient hyperaccumulator plants, it is essential that the TF and BF are greater than 1(Willscher *et al.*, 2013; Ahmadpour *et al.*, 2014).The present study showed that *P. fluorescens* and Ethylene diamine tetraacetic acidapplication significantly increased the B.F and T.F of F treated plants in comparison to control plants. This study concludes that the F removal efficacy of plant treated with *P. fluorescens* and Ethylene diamine tetraacetic acid was higher at lab scale (84.73%)than field study (67.7%) under realistic conditions (Chaudhary and Khan, 2016). The significantly positive correlations were found between F concentrations with the amount of P.F and EDTA in plant organs. This information on field scale and simultaneously with its wide geographical range area suggests that *P. juliflora* could be used for phytoremediation of F contaminated sites under field scale.

4.22 MINERAL CONTENTS IN PLANT ORGANS (ROOT, SHOOT AND LEAVES)AFTER TREATMENT WITH *PSEUDOMONAS FLUORESCENS* AND ETHYLENE DIAMINE TETRAACETIC ACID IN THE FIELD EXPERIMENT

The plant nutrients were significantly different (p=1.09) among the different treatment concentrations of F with microbes and chelates. The macronutrient showed that highest N, P, K contentin roots was found to be higher (50%), while in comparison to control. The level of N, P and K were affected by different concentrations of F(25, 50, 75 and 100 mg kg^{-1}). Therefore macronutrients (N, P and K) enhanced when were given the treatment of P.F due to the biosynthesis of enzymes which have been supportive under stress (Table 4.24.9 and 4.25.9). When the concentrations of chromium were increased from 10-40 mg kg-1 in the soil, consequently macronutrients and micronutrients in shoots (N, P, K and Fe) of *Brassica juncea* were decreased. The K content in the shoot of white lupine (*Lupinus albus*) was significantly decreased in the case of different Cd concentrations 18 and 45 μM (Sharma and Pant, 1994). The amount of micronutrients in roots such as Fe, Mn, Zn and Cu was significantly increased (p=1.56) with the given treatment of P.F and EDTA. The increasing concentrations of macronutrients (N, P and K) were found due to the positive effects of plant growth- promoting bacteria under heavy metal stress (Dobbelaere *et al.*,2003). The micronutrient ranges were analyzed separately which were enhanced approximately 60% in the root (Figure 4.27 and 4.33).

However, the macronutrients in the shoot increased 50% significantly (p=0.98). The ranges of available Fe, Mn, Zn and Cu enhanced approximately 60% in the shoot. While, we found a low range of macronutrients (N, P and K) in leaves approximately 47% as in comparison with shoot and root due to F negative effects.

The low ranges of available micronutrients (Fe, Mn, Zn and Cu) were found to be 50% as shown in Figure 4.27 and 4.33. There had been the negative effect of heavy metals such as Cd, Hg, Cu and As on macronutrients (N, P and K) and micronutrients Fe, Mn, Zn and Cu (Hossain *et al.*, 2014).

Consequently, it may be the possibility that an active pathway and channels interfere with the action of P.F and EDTA; therefore, there is an increase in mineral content in field condition.

The relationship between F and other elements in plants was

investigated under the given treatment of P.F and EDTA. Macronutrients and micronutrients were highly increased due to P.F and EDTA, which have stimulated the solubility of other elements with F in soil.

The positive correlation between F and N, P, K, Fe, Mn, Zn and Cu was observed in the plant. Our research showed that a high amount of F accumulation in the plant was accompanied by high concentrations of elements such as N, P, K, Fe, Mn, Zn and Cu in soil.

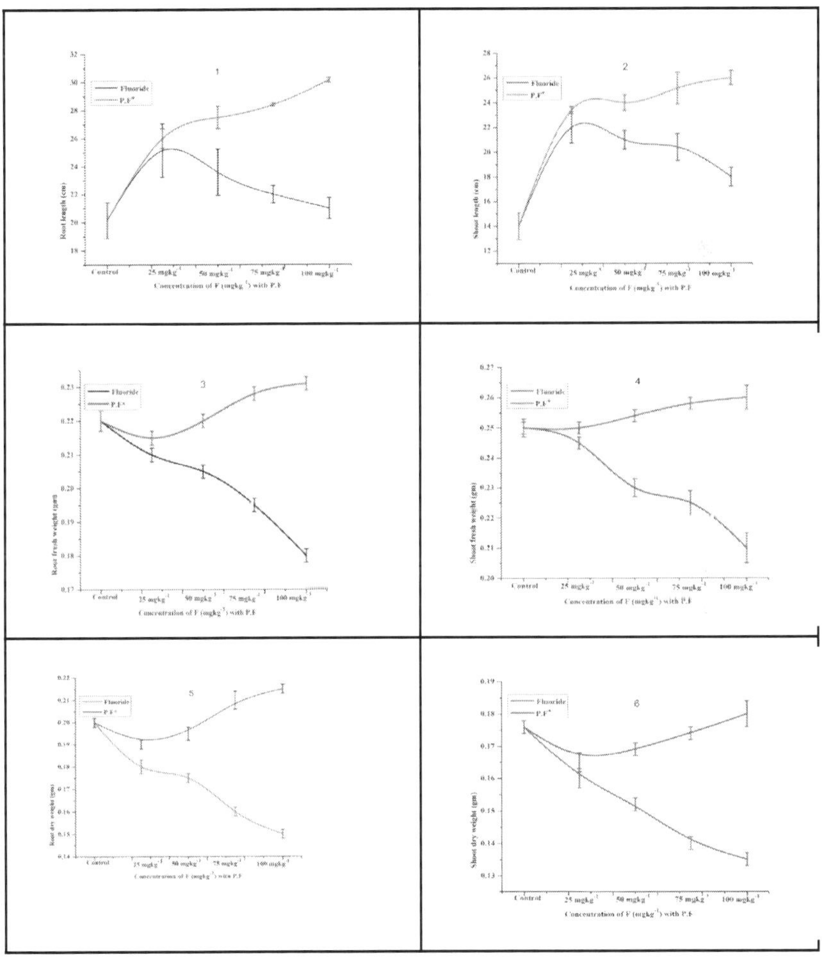

Figure 4.22: Effect of *Pseudomonas fluorescens* (P.F) on plant growth and biomass in the field for 120 days (1-root length, 2-shoot length,3-root dry weight,4-shoot dry weight, 5-root fresh weight, 6-shoot fresh weight and significant value*).

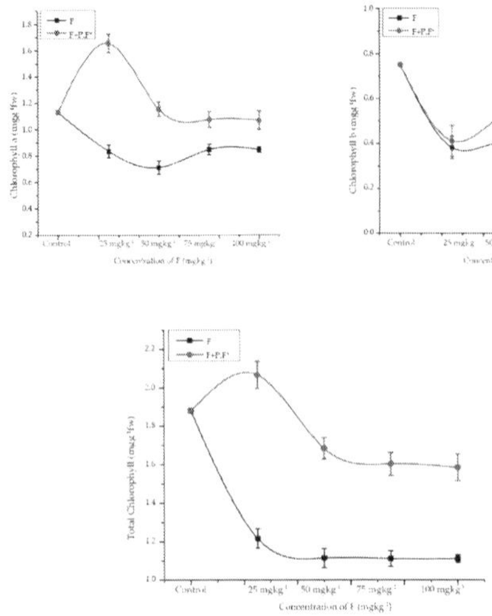

Figure 4.23: Effects of *Pseudomonas fluorescens*(P.F) on chlorophyll content of *P. juliflora* plant in field for 120 days (significant value*).

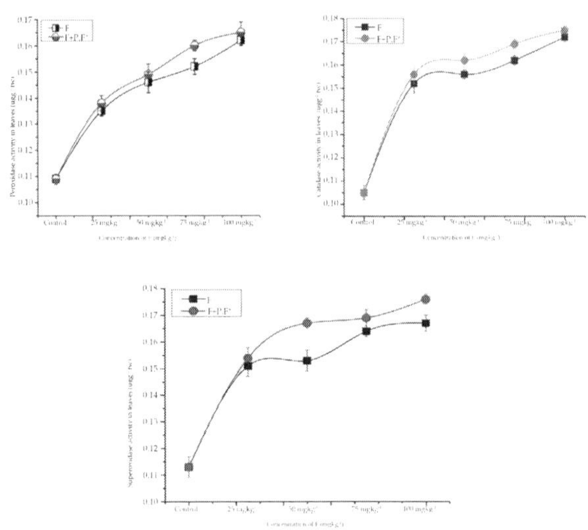

Figure 4.24: Effect of *Pseudomonas fluorescens* (P.F) on antioxidant activity of *P. juliflora* plant in field for 120 days (significant value*).

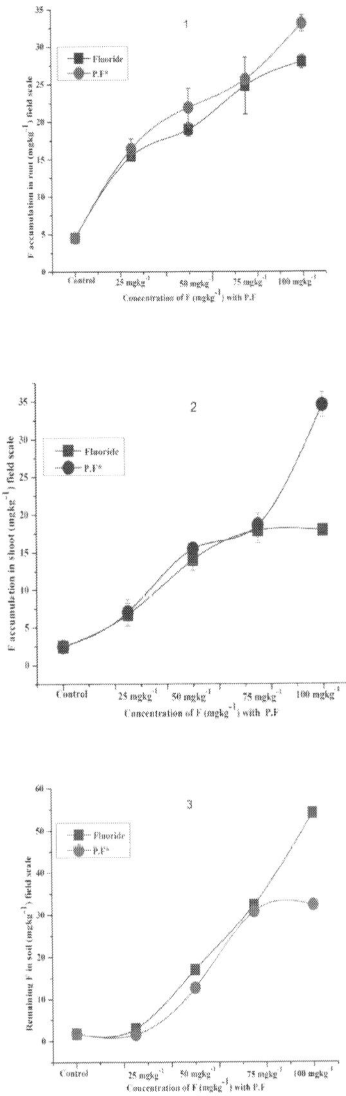

Figure 4.25: Effect of *Pseudomonas fluorescens* (P.F) on *P. juliflora* plant F accumulation in the field for 120 days 1-root uptake, 2-shoot uptake and 3-remaining F in soil (significant value*).

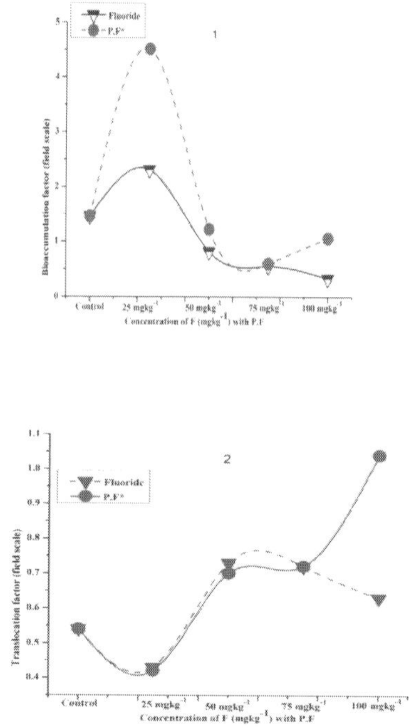

Figure 4.26: Effect of *Pseudomonas fluorescens*(P.F) on bioaccumulation and translocation factor of *P. juliflora* plant in the field for 120 days (significant value*)

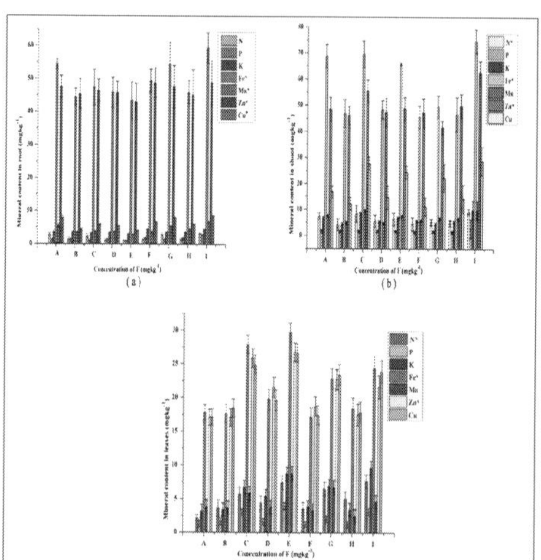

Figure 4.27: Effect of *Pseudomonas fluorescens* (P.F) on root (1), shoot (2), leaves (3) mineral contents of *P. juliflora* for 120 days. A-control, B-25

mgkg-1, C-25 mg kg-1NaF+ P.F, D-50 mg kg-1NaF, E-50 mg kg-1NaF +P.F, F-75 mg kg-1 NaF, G-75 mg kg-1NaF +P.F, H-100 mg kg-1 NaF, I-100 mg kg-1NaF +P.F, (significant value*).

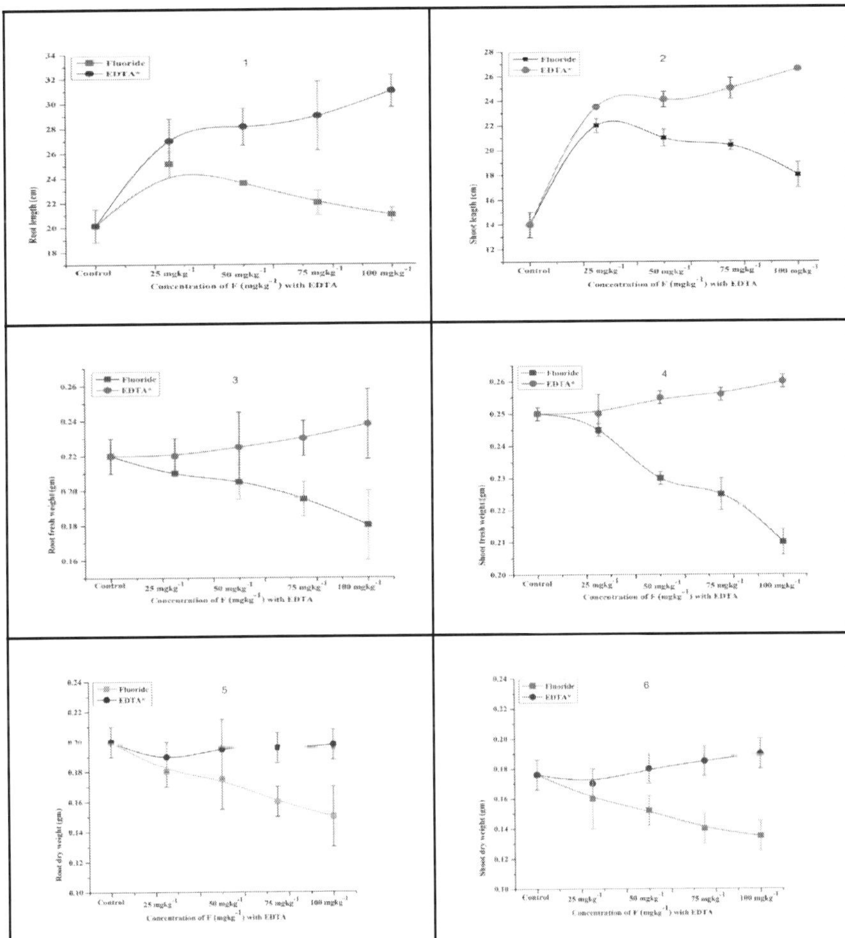

Figure 4.28: Effect of Ethylene diamine tetraacetic acid(EDTA) on plant growth and biomass in the field for 120 days (1-root length, 2-shoot length,3-root dry weight, 4-shoot dry weight, 5-root fresh weight, 6-shoot fresh weight and significant value*).

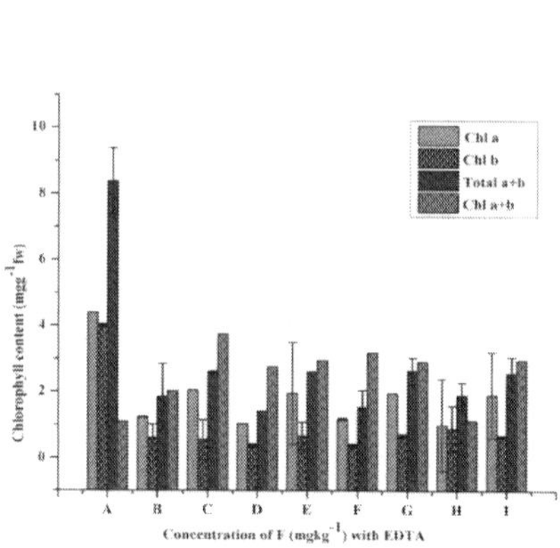

Figure 4.29: Effects of Ethylene diamine tetraacetic acid(EDTA) on chlorophyll content of *P. juliflora* plant in the field for 120 days. A-control, B-25 mg kg^{-1} NaF, C-25 mg kg^{-1}NaF + EDTA, D-50 mg kg^{-1}NaF, E-50 mg kg^{-1}NaF +EDTA, F-75 mg kg^{-1} NaF, G-75 mg kg^{-1}NaF +EDTA, H-100 mg kg^{-1} NaF, I-100 mg kg^{-1} NaF +EDTA (significant value*).

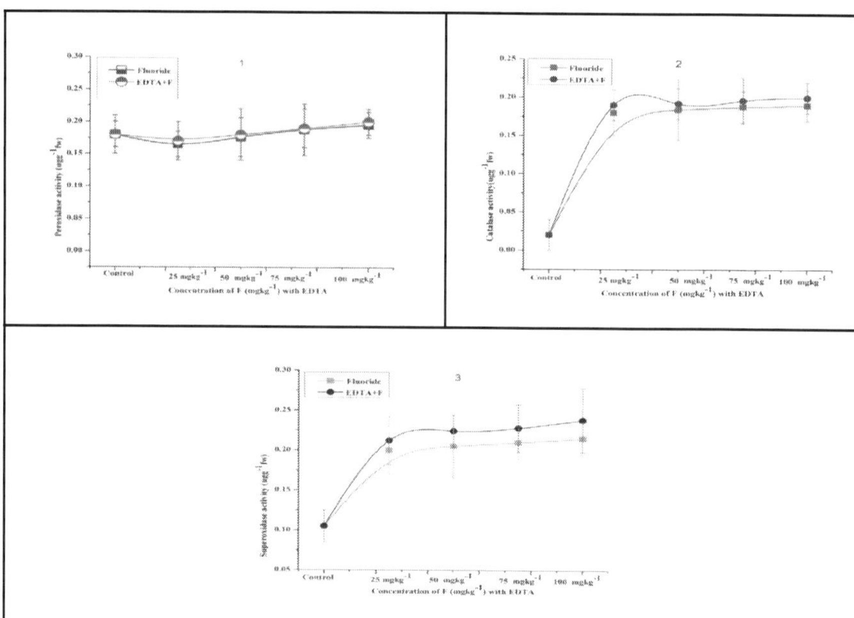

Figure 4.30: Effects of Ethylene diamine tetraacetic acid(EDTA) on antioxidant activity of *P. juliflora* plant in the field for 120 days (significant value*).

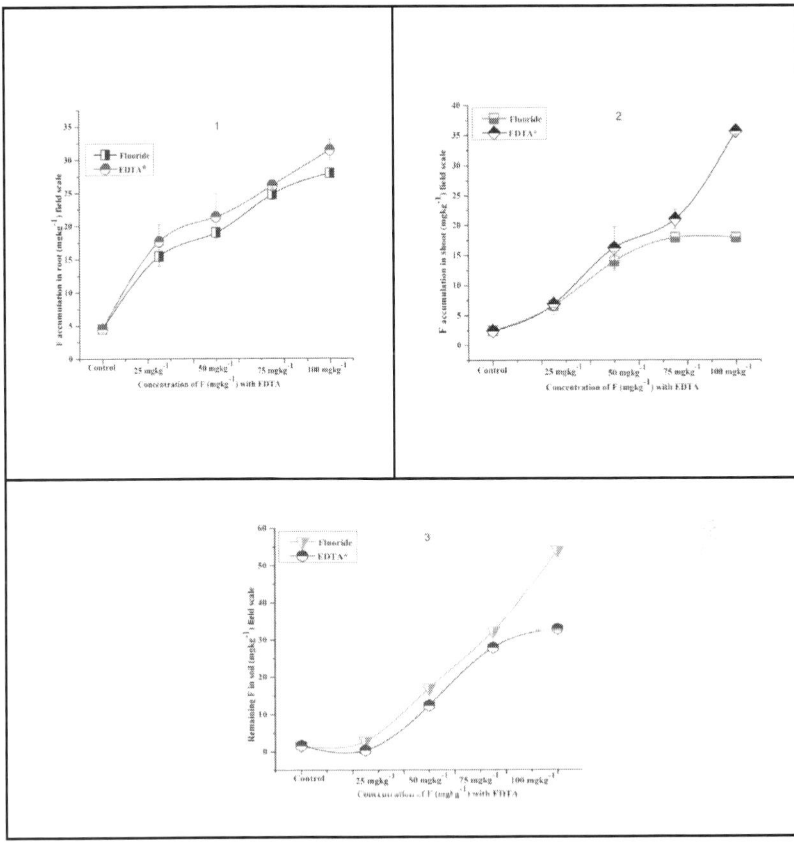

Figure 4.31: Effects of Ethylene diamine tetraacetic acid (EDTA) on *P. juliflora* plant accumulation in the field for 120 days, 1-root uptake, 2-shoot uptake and 3-remaining F in soil (significant value*).

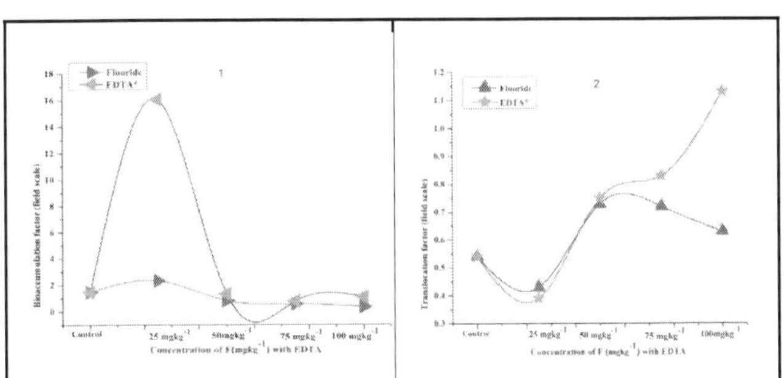

Figure 4.32: Effects of Ethylene diamine tetraacetic acid(EDTA) on bioaccumulation and translocation factor of *P. juliflora* plant in the field for 120 days (significant value*)

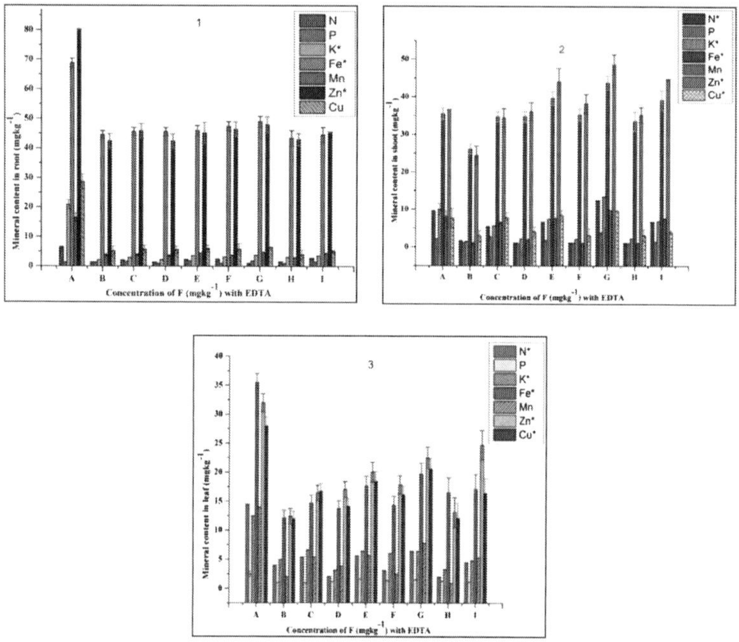

Figure 4.33: Effect of Ethylene diamine tetraacetic acid (EDTA) on root (1), shoot (2), leave (3) mineral contents of *P. juliflora* plant in field for 120 days- A-control, B-25 mg kg^{-1}NaF, C-25 mg kg^{-1}NaF + EDTA, D-50 mg kg^{-1}NaF, E-50 mg kg^{-1}NaF +EDTA, F-75 mg kg^{-1}NaF, G-75 mg kg^{-1}NaF +EDTA, H-100 mg kg^{-1}NaF, I-100 mg kg^{-1}NaF +EDTA (significant value*).

4.23 MINERAL CONTENTS IN SOIL TREATMENT WITH *PSEUDOMONAS FLUORESCENS* AND ETHYLENE DIAMINE TETRAACETIC ACID IN THE FIELD EXPERIMENT

The macronutrient contents have been shown in Table 4.21.2, the highest N, P, K content in their soil was found to be more than (50%), while in comparison to control. The level of N, P and K were affected by different concentrations of F(25, 50, 75 and 100 mg kg^{-1}) in soil. Therefore macronutrients (N, P and K) level enhanced when given treatment of P.F and EDTA due to the mobility increased by microbes and chelates which has been supportive under stress (Tables 4.21.1, 4.21.2 and 4.22.1, 4.22.2). The amount of micronutrients in the soil such as Fe, Mn, Zn and Cu significantly increased with the given treatment of P.F and EDTA.

Table 4.21.1: Mineral content in soil of *P. juliflora* seedlings were grown in soil for 120 days in different concentrations of F viz. control, 25, 50, 75 and 100 mg kg-1 NaF before treatment with *Pseudomonas fluorescens* in the field experiment

Mineral contents in the soil before treatment									
S.No	**pH**	**E.C**	**N**	**P**	**K**	**Fe**	**Mn**	**Zn**	**Cu**
T0	8.52±0.01	1.23±0.00	112±1.51	5.21±1.22	145±0.00	2.53±0.56	2.35±0.56	1.74±0.56	1.12±0.00
T1	7.51±0.00	1.95±0.10	116±0.44	4.62±1.00	108±0.00	2.11±0.56	1.73±0.67	1.73±0.05	2.09±0.00
T2	8.43±0.01	1.05±0.00	114±0.00	5.11±2.55	115±1.55	1.22±0.56	1.23±0.44	2.91±0.55	2.57±1.55
T3	7.25±0.01	1.71±0.10	115±0.00	4.50±0.45	106±0.56	1.73±0.65	1.76±1.56	1.72±0.23	1.08±0.56
T4	8.65±0.02	1.45±0.03	115±1.10	5.51±0.33	113±0.55	1.44±0.67	1.28±1.11	1.83±0.00	1.66±0.65
T5	7.63±0.02	1.81±0.45	114±0.22	3.43±0.65	118±1.55	1.56±0.77	1.77±0.55	1.67±0.00	1.67±0.65
T6	7.81±0.10	1.45±0.05	115±0.42	5.71±0.23	123±0.65	1.12±1.55	1.35±0.67	1.66±0.55	1.16±0.67
T7	8.23±1.22	1.62±0.00	113±0.34	3.61±0.56	129±1.55	1.27±0.45	1.23±0.90	1.33±0.56	1.15±0.78
T8	8.78±0.01	1.45±0.33	115±0.89	6.40±0.56	114±1.66	1.29±1.45	2.17±0.56	2.41±0.50	1.54±0.66

Table 4.21.2: Mineral content in soil ofP. *juliflora* seedlings were grown in soil for 120 daysin different concentration of F viz. control, 25, 50, 75 and 100 mg kg-1 NaF after treatment with *Pseudomonas fluorescens* in field experiment

Mineral contents in soil after treatment									
S.No	**pH**	**E.C**	**N**	**P**	**K**	**Fe**	**Mn**	**Zn**	**Cu**
T0	6.41±0.03	1.44±0.05	135±1.25	7.70±1.55	146±0.65	2.75±0.30	2.69±0.54	2.77±0.60	2.08±0.50

T1	6.28±0.04	1.43±0.04	112±1.00	4.71±0.50	158±4.55	3.56±0.50	3.23±0.60	1.23±0.70	1.18±0.50
T2	6.23±0.05	1.28±0.01	141±1.56	6.71±0.70	168±4.00	3.56±0.60	2.77±0.70	1.67±0.50	1.07±0.00
T3	7.28±1.00	1.54±0.05	113±1.00	4.62±0.50	139±5.00	4.45±0.70	3.23±0.80	1.33±0.40	2.15±0.00
T4	6.81±0.00	1.24±0.05	142±1.55	6.73±0.40	176±0.50	3.55±0.60	2.87±0.50	3.76±0.60	1.08±0.00
T5	7.41±0.04	1.44±0.06	125±2.88	4.72±0.60	146±0.56	2.75±0.40	2.69±0.56	2.77±0.40	2.08±0.06
T6	6.30±0.02	1.26±0.56	142±1.87	7.50±0.70	179±0.75	3.95±0.60	2.94±0.60	3.87±0.45	2.06±0.70
T7	7.43±0.01	1.74±0.05	143±1.55	3.22±0.65	146±1.45	3.43±0.70	2.54±0.70	2.65±0.78	1.05±0.80
T8	6.02±0.04	1.04±0.60	144±2.50	9.91±0.70	180±4.55	4.53±0.60	3.54±0.80	4.65±0.80	2.65±0.60

Individual readings n =3 for each plant were averaged and are presented with ±SD. Significant correlation was found at 0.05 level of significance. Individual readings n =3 for each plant were averaged and are presented with ±SD. Significant correlation was found at 0.05 level of significance. T0-control, T1-25 mg kg^{-1} NaF, T2-25 mg kg^{-1} NaF +P.F, T3- 50 mg kg^{-1} NaF, T4-50 mg kg^{-1} NaF +P.F, T5-75 mg kg^{-1} NaF, T6-75 mg kg^{-1} NaF +P.F, T7- 100 mg **kg^{-1} NaF, T8-** 100 mg kg^{-1} NaF + P.F. P.F: Pseudomonas fluorescens,

Table 4.22.1: Mineral content in soil of *P. juliflora* seedlings were grown in soil for 120 days in different concentrations of F viz. control, 25, 50, 75 and 100 mg kg-1 NaF before treatment with 20 mM kg^{-1}Ethylene diamine tetraacetic acid in field experiments

Mineral contents in the soil before treatment									
S.No	**pH**	**E.C**	**N**	**P**	**K**	**Fe**	**Mn**	**Zn**	**Cu**
T0	7.52±0.02	1.38±0.05	112±0.02	10.20±0.02	123±0.03	2.53±0.02	2.35±0.03	4.74±0.01	3.12±0.55
T1	7.51±0.03	1.39±0.01	116±4.04	10.62±0.03	108±0.02	1.11±0.02	1.73±0.05	1.73±0.22	2.09±0.34

T2	7.27±0.05	1.43±0.04	114±2.00	14.91±0.02	125±0.03	1.10±0.01	2.12±0.01	3.71±0.43	3.12±0.44
T3	8.25±0.06	1.87±0.02	115±2.00	14.51±0.02	114±0.04	1.73±0.02	1.76±0.00	1.72±0.00	2.18±0.23
T4	7.42±0.02	1.43±0.01	115±7.01	16.43±0.01	133±0.05	1.73±0.02	4.23±0.01	3.73±0.00	3.14±1.22
T5	7.63±0.02	1.58±0.02	114±1.02	15.42±0.02	128±0.02	1.86±0.01	1.77±0.02	1.77±0.23	2.67±0.12
T6	8.20±0.00	1.76±0.03	94±2.02	16.51±0.02	141±0.01	1.85±0.02	4.44±0.03	1.67±0.34	3.06±2.55
T7	8.23±0.03	1.26±0.03	113±4.54	15.61±0.03	139±0.02	2.27±0.01	2.23±0.01	3.33±0.54	2.75±0.55
T8	7.12±0.02	1.74±0.03	93±2.84	16.21±0.01	156±0.04	2.43±0.02	4.54±0.02	4.69±0.66	3.95±0.06

Table 4.22.2: Mineral content in soil of *P. juliflora* seedlings were grown in soil for 120 days in different concentrations of F viz. control, 25, 50, 75 and 100 mg kg-1 NaF after treatment with 20 mM kg^{-1}Ethylene diamine tetraacetic acid in the field experiment

Mineral contents in the soil after treatment									
S.No	**pH**	**E.C**	**N**	**P**	**K**	**Fe**	**Mn**	**Zn**	**Cu**
T0	7.41±0.00	1.44±0.05	145±5.00	14.7±2.55	146±0.05	2.75±0.05	2.69±0.50	2.77±0.01	2.08±0.00
T1	6.28±0.00	1.43±0.00	112±4.00	12.7±3.55	158±0.06	3.56±0.02	3.23±0.00	5.23±0.03	1.18±0.03
T2	6.77±0.00	1.46±0.00	94±2.00	15.5±4.00	141±1.55	2.45±0.03	2.74±0.00	2.55±0.01	3.06±0.05
T3	7.28±0.00	1.54±0.00	113±5.00	12.6±3.00	139±1.00	4.45±0.05	3.23±0.50	4.33±0.50	2.15±0.04
T4	6.78±0.00	1.54±0.00	193±4.55	16.2±3.00	146±2.44	2.53±0.04	2.78±0.60	2.55±0.55	3.05±0.02
T5	7.41±0.00	1.44±0.50	195±3.55	16.7=2.00	146±1.55	2.75±0.06	2.69±0.70	2.77±0.55	2.08±0.00
T6	6.25±0.00	1.14±0.60	193±5.22	17.6=2.00	179±1.00	4.27±0.05	3.23±0.50	4.33±0.50	3.15±0.03
T7	7.43±0.00	1.74±0.06	143±4.22	16.2=1.50	146±1.55	3.43±0.06	2.54±0.40	4.65±0.55	1.05±0.04
T8	6.53±0.00	1.14±0.00	195±5.00	18.7=1.55	173±2.00	5.12±0.40	4.35±2.00	6.66±1.00	6.16±0.50

Individual readings n =3 for each plant were averaged and are presented with ±SD. Significant correlation was found at 0.05 level of significance. T0-control, T1-25 mg kg^{-1} NaF, T2-25 mg kg^{-1} NaF +EDTA, T3- 50 mg kg^{-1} NaF, T4-50 mg kg^{-1} NaF +EDTA, T5-75 mg kg^{-1}NaF, T6-75 mg kg^{-1} NaF +EDTA, T7- 100 mg kg^{-1} NaF, T8-100 mg kg^{-1}NaF + EDTA. EDTA: Ethylene diamine tetraacetic acid.

Fluoride solubility may be associated with the solubility of another element with which it forms complexes. The F has a positive correlation with another element. It was found to have the highest significant correlation with Fe, Mn, Zn and Cu (Hong *et al*., 2016). A significant positive correlation coefficient (r) values were observed between F and pH and also other elements. The positive highest correlations were shown between F and C, Fe and Cu (r =0.57, 0.70, 0.81, 0.50, 0.76, 0.68 and r = 0.72, 0.92, 0.78, 0.85, 0.56, 0.67) in the contaminated soil before and after given treatment of P.F and EDTA.Fluoride mobility in soil is dependent on the pH and the types of sorbents present. Fluoride was strongly bonded with Fe, Al, and Ca, Mg, Mn and Cu (Omueti and Jones, 1977) and its mobility also depends on the amount of water percolating into the soil zone (Abugri, 2010). The F concentration was increased in the soil which might be generally due to the release of OH^- throughout the adsorption process (Bower and Hatcher, 1967). The F solubility is controlled through F adsorption by inorganic constituents of the acidic soil (Loganathan *et al*., 2003; Saxena and Rani, 2012).

We reported that F uptake efficiency increased of hyperaccumulator plant *P. juliflora* through the application of *P. fluorescens* and EDTA both in laboratory and field scale. Application of *P. fluorescens* and EDTA enhanced the F removal efficiency of *P. juliflora*. The present study not just removed the F contaminants from the soil (groundwater) but also improved the soil property for further use in agricultural purposes. The results concluded that *P. fluorescens* and EDTA application not only enhanced the F uptake but also increased the biomass content of plants due to the availability of minerals in the soil. Such an application can be used to analyze the potential for F contaminated soil in semi- arid zones to ensure sustainable groundwater utilization. We believe that it could be providing an efficient and environmentally friendly solution for F remediation. In addition, this study can be helpful or improve land management for agriculture purposes.

5

Summary and Conclusion

Fluoride (F) contamination is one of the major issues that humanity is facing in the twenty-first century. F contamination in groundwater has now become one of the most important health- related problems in India and many other regions of the world. It is a global environmental problem that affects crop yields, soil biomass and leads to bioaccumulation of F in the food chain. There are specific physiological processes that are known to be markedly affected by F including decreased growth, chlorosis, and leaf tip burn and leaf necrosis, mineral contents in plants. Accumulation in vegetables, particularly in areas irrigated and non-irrigated soils with highly F contaminated water may pose a direct threat to human health and cause diseases such as hypertension and fluorosis. Some amount of F is essentially required for the human body for healthy bones and teeth. However, when its accumulation or absorption crosses the recommended limit of WHO and BIS i.e 1.5 mg/l, it leads to mild dental fluorosis and to crippling skeletal fluorosis as the quantity and period of exposure increases. A new cleanup technology termed phytoremediation is an environment- friendly green solution involving living plants which offers a cost -effective means for cleaning F contaminated soils.

The present study reported that a combination of geo-statistical andmultivariate statistical analysis to examine the spatial distribution of Fluoride contamination in soil is valuable and good for public health protection. This work also studies the soil quality parameters while performing spatial studies. It is used to carve out spatial distribution maps of the study area covering seven sites of Banasthali Newai Tehsil, Tonk, Rajasthan, India. Proper legends developed over each spatial map to describe the relative distribution of each quality parameters. It can be concluded from this work that soil is not suitable for agriculture purpose of the studied area.

This study also identifies the F-tolerant bacteria *Clostridium sp.* which improves the phytoremediation efficiency and biomass of *P. juliflora*. Microbe-assisted phytoremediation technology helps in the development of plants with a higher potential to clean our environment. *Clostridium sp.* KP136288-FTB1, KP136289-FTB1, KP136290-FTB2 and KP136291-FTB2 showed significant results under F stress. The PGPR effect of the bacterial strains with high F tolerance was found in the current study. This study can be further utilized for commercialization. Secondly, in the present study, two methods have been used to treat F contamination in soil. These two methods of F treatment are microbe and chelate assisted phytoremediation for lab and field scale. These methods after applying in the agriculture field gave satisfactory results. *Pseudomonas fluorescens*(P.F) improves the abilities of *P. juliflora* phytoremediation efficiency by increasing plant biomass. F had a negative impact on chlorophyll pigments of *P. juliflora*. Fluoride has high interactions especially at its high concentrations with mineral content in plant especially P, Fe and Cu, which led to decrease mineral levels in *P. juliflora*. Based on the present study, it can be concluded that soil nutrients contents are sensitive to F concentration.*P. juliflora* need the microbial treatment in the soil for improving F remediation. However, this work could be commercialized using the microbial treatment. Therefore, the P.F could be useful for remediation purposes, counterbalance nutrients content and sustainable agriculture especially in the areas where F contaminated problems occurred in soil and water.

Chelating agent EDTA improves the F removal capacity in soil. Thus among all the concentrations, at 120 days *P. juliflora* plant removes F to a greater extent in comparison to 30, 60, and 90 days plants. A significant correlation was found between translocation (TF) and bioaccumulation factor (B.F) with an increase in F concentration with EDTA. The more F accumulation in the shoots, which combined with TFs and BFs, indicate the potential of *P. juliflora* for field applications. The major decrease of various plant minerals (N, P, K and Fe, Mn, Zn and Cu)was seen at 100 mg kg^{-1} NaF. In addition, F accumulation in root and shoot of *P.juliflora* increased by P.F and EDTA. The result suggested that plants subjected to P.F and EDTA treatment levels had a good effect on growth, chlorophyll, root length, root dry weight, shoot length, and shoot dry weight, macronutrients and micronutrients under Fcontaminated soil at 100 mg kg^{-1}.

This study concludes that the P.F and EDTAhas significantly increased the translocation of F from roots to shoots and increased F uptake in *P. juliflora* by more than 50%in pot and field conditions. *P. juliflora* is a good candidate to remove F due to its high shoot biomass when grown in F contaminated soil with 100 mg kg^{-1}NaF and its capacity of tolerance. Fluoride accumulation in the root of this plant is very satisfactory due to its large root depth in the soil. Therefore, P.F and EDTA is an appropriate agent to effectively enhance the phytoextraction of F by *P. juliflora* from F contaminated soil. Meanwhile, *P. juliflora* plants have high biomass, high economic value and it poses no danger of transferring F into the food chain.F contamination could be effectively reduced by applying new phytoremediation approaches involving P.F and EDTA both at laboratory and field scale.

As an outcome of this study, the harvested plant biomass of *P. juliflora* from phytoremediation of the test field site can be utilized further and the helpful combination could be an innovative method of remediation techniques. The *P. juliflora* plant in the present study showed a substantial improvement in the ability to accumulate F. Earlier it was reported that *P. juliflora* plant removes F up to 34.13%, but the present study improved this procedure.

Application of P.F improved the removal efficiency by *P.juliflora* plant by 84.73% in the pot and 67.70% in the field. Chelating agent EDTA improves phytoremediation efficiency by 68.38% in the pot and 67.12% in the field. The study not only the F contaminants have been removed from the soil but also the soil property has been improved for further use in agricultural purposes and recycling the biomass for various applications. The present study improved phytoremediation technology could be easily commercialized for the treatment of F contaminated lands and water.

References

1. Abdul, M.R., Schroder, P. (2009). Implications of metal accumulation mechanisms to phytoremediation. Environmental Science and Pollution Research, **16,** 162-175.
2. Abollino, O., Giacomino, A., Malandrino, M., Mentasti, E., Aceto, M., Barberis, R. (2006). Assessment of metal availability in a contaminated soil by sequential extraction. Water Air Soil Pollution,**173(1-4),**315-338.
3. Abou-Shanab, R.A.I., Angle, J.S., Chaney, R.L. (2006). Bacterial inoculants affecting nickel uptake by *Alyssum murale* from low, moderate and high Ni soils. Soil Biology and Biochemistry **38,** 2882-2889.
4. Abugri, D.A., Pelig-Ba, K.B. (2011). Assessment of F content in tropical surface soils used for crop cultivation. African Journal of Environmental Science and Technology,**5(9),** 653-660.
5. Agalakova, N.I., Gusev, G.P. (2012). Molecular mechanism of cytotoxicity and apoptosis induced by inorganic fluoride, ISRN Cell Biology, 16.
6. Agarwal, V., Vaish, A. K., and Vaish, P. (1997). Groundwater quality on fluoride and fluorosis in Rajasthan. Current Science,**73,** 743-64.
7. Ahmadpour, P.,Ahmadpour, F., Mahmud, T., Abdu, A., Soleimani, M., Tayefeh, F.H.(2014). Phytoremediation of heavy metals: a green technology. *In:* Asrari, E. (Ed), heavy metal contamination of water and soil: Analysis, Assessment, and Remediation strategies. CRC Press, Taylor and Francis, 249.
8. Akhter, M.F., Omelon, C.R., Gordon, R.A., Moser, D., Macfie, S.M.

(2014). Localization and chemical speciation of cadmium in the roots of barley and lettuce. Environmental Experimental Botany,**100,** 10-19.

9. Allen,S.C. (1989). Chemical analysis of ecological material. ed. S.E Stewars, Blackman Scientific Oxford, 368.
10. Ali,M.B., Vajpayee, P., Tripathi, R.D., Rai, U.N., Singh, S.N., Singh, S.P. (2003). Phytoremediation of lead, nickel and copper by *Salix acmophylla Boiss*; Role of antioxidant enzymes and antioxidant substances. Archives of Environmental Contamination and Toxicology,**70**, 462-469.
11. Ali, H., Khan, E., Sajad, M.A. (2013). Phytoremediation of heavy metals-concepts and applications. Chemosphere,**91(7),** 869-881.
12. AL-Qurainy,F., Abdel-Megeed, A. (2009). Phytoremediation and detoxification of two organophosphorus pesticides residues in Riyadh area. World Application Science Journal, **6(7),** 987-998.
13. Anandham, R., Indiragandhi, P., Madhaiyan, M., Ryu, K.Y., Jee, H.J., Sa, T.M.(2008). Chemolithoautotrophic oxidation of thiosulfate and phylogenetic distribution of sulfur oxidation gene (soxB) in rhizobacteria isolated from crop plants. Research Microbiology,**159,** 579-589.
14. Arnon, D. I. (1949). Copper enzyme in isolated chloroplast. Polyphenol oxidase in *Beta vulgaris*. Plant and Soil,**191,** 13-25.
15. Ashok, K., Bisht, B.S., Manish, K., Lalit, K.(2010). Effects of Ni and Zn on growth of vigna mungo, Vigna radiata and glycine max. International Journal of Pharmacy and Biological Sciences,**1,** 2.
16. Ashraf, A., Bibi, I., Niazi, N.K., Ok, Y.S., Murtaza, G., Shahid, M., Kunhikrishnan,A., Mahmood, T. (2016). Chromium(VI) immobilisation efficiency of acid-1 activated banana peel over organo-montmorillonite in aquatic environments. International Journal of Phytoremediation,http://dx.doi.org/10.1080/ 15226514.2016.1256372.
17. Azcón, R., del Carmen P.M., Biró, B., Roldán, A., Ruíz-Lozano, J.M. (2009). Antioxidant activities and metal acquisition in mycorrhizal plants growing in a heavy-metal multicontaminated soil amended with treated lignocellulosic agrowaste. Applied Soil Ecology, **41(2),** 168-177.
18. Azhar, P., Jacob, A., Christopher, C., Nishma, A., Vasu, D.A., Appanna, D. (2015). Metabolic reconfigurations aimed at the detoxification of a multi-metalstress in *Pseudomonas fluorescens*: Implications for the

bioremediation of metal pollutants. Journal of Biotechnology, **200,**38-43.

19. Badr, A.M., Naoko, E., Chang, S.K., Xiaotao, Bi. (2017). The role of tailored bichor in increasing plant growth and reducing bioavailability, phytoxicity and uptake of heavy metals in contaminated soil. Environmental Pollution,**230,** 329-338.
20. Baker, A.J.M., McGrath, S.P., Reeves, R.D., Smith, J.A.C. (2000). Metal hyperaccumulator plants: a review of the ecology and physiology of a biological resource for phytoremediation of metal polluted soils. *In:* N Terry, G Banuelos (eds.) phytoremediation of contaminated soils and waters. CRC Press LLC, Boca Raton, FL, USA, 85-107.
21. Baloch, M.J., Khan, N.U., Rajput, M.A., Jatai, W.A., Gul, S., Rend I.H., Veesar, N.F. (2014). Yield related morphological measures of short duration genotypes. Journal of Animal Plant Science, **24(4),** 1198-1211.
22. Barabasz, A., Wilkowska, A., Tracz, K., Ruszczyñska, A., Bulska, E., Mills, R.F.(2013). Expression of *HvHMA2*in tobacco modifies Zn-Fe-Cd homeostasis. Journal of Plant Physiology, **170,**1176-1186.
23. Baunthiyal, M. (2008). Screening of local flora for bioremediation of F rich soil and water. Ph.D Thesis Banasthali University, 91-93.
24. Bauddh, K.,Singh, R.P. (2012). Cadmium tolerance and its phytoremediation by two oil yielding plants *Ricinus communis* (L) and *Brassica Juncea* (L) from the contaminated soil. International Journal of Phytoremediation, **14,** 772-785.
25. Beauchamp, C., Fridivich, I. (1971). Superoxide dismutase improved assay and an assay applicable to acrylamide gels. Analysis Biochemistry,**44(1),** 267-87.
26. Belimov, A.A., Hontzeas, N., Safronova, V.I., Demchinskaya, S.V., Piluzza, G.,Bullitta, S., Glick, B.R.(2005). Cadmium-tolerant plant growth promoting bacteria associated with the roots of Indian mustard (Brassica juncea L. Czern). Soil Biology and Biochemistry, **37,** 241-250.
27. Bengdtsson, B., Asp, H., Jensn, P. (1994). Uptake and distribution of calcium and phosphorus in Beech (*Fagus sylvatica*) as influences by aluminum and nitrogen. Tree Physiology,**14(1),** 63-73.
28. Bezel, V. S., Zhuikova, T.V., Pozolotina, V.N. (1998). The structure of dandelionceno populations and specific features of heavy metal

accumulation. Russian Journal of Ecology,**29,** 331-337.

29. Bhargava, B.S., Raghupathi, H.B. (1999). Multivariate diagnosis of nutrient balance in grapes (*Vitis vinifera L*) cv Anab-e-shahi. Journal of Plant Nutrition, **22(2),** 219-227.
30. Bhargava, D., Bhardwaj, N. (2010). Effect of sodium fluoride on seed germination and seedling growth of Triticum aestivum Var. Raj. 4083. Journal of Phytoremediation, **4,** 41-43.
31. Bhardwaj, P., Chaturvedi, A.K., Prasad, P. (2009). Effect of enhanced lead and cadmium in soil on physiological and biochemical attributes of *Phaselousvulgaris L.* Nature Science, **7(8),** 63-75.
32. Biswas, J.K., Mondal, M., Rinklebe, J., Sarkar, S.K., Chaudhuri, P., Rai, M., Shaheen,S.M., Song, H., Rizwan, M. (2017). Multi-metal resistance and plant growth promotion potential of a wastewater bacterium *Pseudomonas aeruginosa* and its synergistic benefits. Environment Geochemistry and Health, doi 10.1007/s10653-017-9950-5.
33. Blaylock, M.J., Salt, D.E., Dushenkov, S., Zakarova, O., Gussman, C., Kapulnik,Y., Ensley , B.D. and Raskin, I. (1997). Enhanced accumulation of Pb in indian mustard by soil- applied chelating Agents. Environmental Science and Technology, **31, 860**-865.
34. Bose, J.,Babourina, O., Shabala, S., Rengel, Z. (2013). Low-pH and aluminium resistance in *Arabidopsis* correlates with high cytosolic magnesium content and increase magnesium uptake by plant roots. Plant Cell Physiology,**54(7),** 1093-1104.
35. Bractic, A.M., Majic, D.B., Samardzc J.T., Maksimovic V.R. (2009). Functional analysis of the buckwheat metallothionein promoter: tissue specificity pattern and up-regulation under complexes stress stimuli. Journal of Plant Physiology,**166,**996-1000.
36. Brandl, H., Faramarzi, M.A. (2006). Microbe metal-interactions for the biotechnological treatment of metal-containing solid waste. China Particuology,**4,** 93-97
37. Brewer, G.J. (1967). Achromatic regions of tetrazolium stained starch gel: inherited electrophoretic variation, American Journal of Human Genetics,**19,** 674-680.
38. Brindha, K., Elango, L. (2011). Fluoride in groundwater causes implications and mitigation measures in: monroy, S.D. (Ed). Fluoride Proper. Application Environment Management, 111-136.

39. Broadhurst, C.L., Chaney, R.L., Angle, J.S., Maugel, T.K., Erbe, E.F., Murphy, C.A. (2004). Simultaneous hyperaccumulation of nickel, manganese, and calcium in *Alyssum* leaf trichomes. Environmental Science and Technology,**38,** 5797-5802.

40. Brooks, R., Lee, J., Reeves, R.D., Jaffre, T. (1977). Detection nickeliferous rocks by analysis of herbarium specimens of indicator plants. Journal of Geochemical Exploration,**7,** 49-57.

41. Cappuccino,J.G., Sherman, N. (1992). Biochemical activities of microorganisms. *In*: Microbiology, A Laboratory Manual. The Benjamin/Cummings Publishing Co. California, USA, 125-179.

42. Carla, P., Nancy, D.M., Laura, I.L., Iustman, J.R. (2012). Isolation and characterization of benzene, toluene and xylene degrading *Pseudomonas* sp. selected as candidates for bioremediation. International Biodeterioration Biodegradation,**67,**15-20.

43. Castric, P.A. (1975). Hydrogen cyanide, a secondary metabolite of *Psuedomonas aeruginosa*. Journal of Microbiology,**21**, 613-618.

44. Castiglione, S., Todeschini, V., Franchin, C., Torrigiani, P.,Gastaldi, D., Cicatelli, A., Rinaudo, C., Berta, G., Biondi, S., Lingua, et al. (2009). Clonal differences in survival capacity, copper and zinc accumulation, and correlation with leaf polyamine levels in poplar: A large-scale field trial on heavily polluted soil. Environmental Pollution,**157,** 2108-2117.

45. Chakarbarti, S., Patra, P.K., Mondal, B. (2013). Uptake of fluoride by two paddy (*Oryza sativa* L.) varieties treated with fluoride contaminated water. Paddy Water Environment, **11,** 619-623.

46. Chae, G.Y., Seong, T.M., Bernhard, K., Kyoung, K., Seong-Yong, K. (2007). Fluoride geochemistry in bed rock groundwater of South Korea. Science of The Total Environment, **385,** 272-283.

47. Chao, J., Pamila, J.B., Brown, A.D., and Yves, V.B. (2013). Sequential evolutionof bacterial morphology by cooperation of a developmental regulator. Nature,**506,** 489-493.

48. Changcheng, L., Yuguo, L., Ke, G., Dayong, F., Guoqing, Li., Yuanrun, Z., Lifei, Yu, Rui, Y. (2012). Effect of drought on pigments, osmotic adjustment and antioxidant enzymes in six woody plant species in Karst habitats of southwestern China. Environmental and Experimental Botany,**71,** 174-183.

49. Chatterjee, J., Chatterjee, C. (2000). Phototoxicity of cobalt, chromium and copper in cauliflower. Environmental Pollution, **109(1),** 69-74.

50. Cheng, K.L., Bray, R.H. (1995). Determination of calcium and magnesium in soil and plant material. Soil Sciences, **72,** 449-458.

51. Chaudhary, K., Khan, S. (2016). Physicochemical characterization of Fluoride (F) contaminated soil and its microbe-assited bioremediation by *Prosopis juliflora*. J. Plant Biology and Soil Health, **3(2),** 8.

52. Chaudhary, K., Sumira, J., Khan, S.(2016). Heavy metal ATPase (*HMA2, HMA3*

& HMA4) genes in hyperaccumulation mechanism of heavy metals. Plant metal interaction (Emerging Remediation Techniques), 545-552.

53. Chaudhary, K., Agarwal S., Khan, S. (2018).Role of phytochelatins (PCs), metallothionines (MTs) and heavy metal ATPase (HMA) genes in heavy metal tolerance. R. Prasad (ed.), mycoremediation and environmental sustainability, fungal biology, https://doi.org/10.1007/978-3-319-77386-5_2 39-60 Springer.

54. Chaudhary, K., Khan, S. (2018).Role of plant growth promoting bacteria (PGPB) for bioremediation of heavy metals: an overview. (biostimulation remediation technologies for groundwater contaminants-IGI-GLOBAL).

55. Chugh, V., Kaur, N., Gupta, A.K. (2011). Evaluation of oxidative stress tolerance in maize (*Zea mays* L.) seedlings in response to drought. Indian Journal of Biochemistry and Biophysics,**48,** 47-53.

56. Cumming, J.R., Eckert, R.T., Evans, L.S. (1986). Effect of aluminium on ^{32}P uptake and translocation by red spruce seedlings. Canadian Journal ForForest Research,**16(4),** 864-867.

57. Curl,A.E., Truelove, B. (1986). The Rhizosphere. Springer-Verlag, New York.

58. Daldoum,DM.A., Musa, S.M. (2012). Effects of Mesquite (*Prosopis juliflora,* Swartz, DC.)manure and NPK fertilizer on growth and yield of forage sorghum (*Sorghum bicolor* L.). Agricultural Research Review, **1(5),**174-179.

59. Daniel, C.P., Volk, V.V. (1985). Fluoride sorption and desorption in soils. Soil Science Society of America Journal,**49,** 583-586.

60. Das, B., Talukdar, J., Sarma, S., Gohain, B., Dutta, R.K., Das H.B., Das, S.C. (2003). Fluoride and other constitutents in groundwater of Guwahati, Assam, India Current Science, **85(5),** 657-661.

61. Datt, J.K., Maitra, A., Mondal, N.K., Banerjee, A. (2012). Studies on the impact of F on germination and seedling growth of gram seed (*Cicer arietinum* L. cv.Anuradha). Journal of Stress Physiology Biochemistry,**8,** 194-202.

62. Datta, P.S., Tyagi, S.K., Mookerjee, P., Bhattacharya, S.K., Gupta, N., Bhatnagar,P.D. (1999). Groundwater NO_3 and fluoride contamination processes in Pushkar valley Rajasthan as reflected from ^{18}O isotope signature and ^{3}H recharge studies. Environment Monitoring Assessment,**56,** 209-219.

63. Datta, P.S., Deb, D.L., Tyagi, S.K. (1996). Stable isotope (180) investigation on the processes controlling fluoride contamination of groundwater. Journal of Contamination Hydrology,**24,** 85-96.

64. Davies, K.L., Davies, M.S., Francis, D. (1991). Zinc induced vaculation in root meristematic cells of *Festuca rubra L.* Plant Cell Environment,**14,** 399-406.

65. Dell'Amico, E., Cavalca, L., Andreoni, V. (2008). Improvement of *Brassica napus* growth under cadmium stress by cadmium-resistant rhizobacteria. Soil Biology and Biochemistry, **40,** 74-84.

66. Desbarats, A.J. (2009). On elevated fluoride and boron concentrations in groundwaters associated with the Lake Saint-Martin impact structure, Manitoba. Applied Geochemistry, **24,** 915-927.

67. Diplock, E.E., Mardlin, D.P., Killham, K.S., Paton, G.I. (2009). Predicting bioremediation of hydrocarbons: Laboratory to field scale. Environmental Pollution,**157,** 1831-1840.

68. Dixit,R., Malaviya, D., Pandiyan, K., Singh, U.B., Sahu, A.(2015). Bioremediation of heavy metals from soil and aquatic environment: an overview of principles and criteria of fundamental processes. Sustainability. **7,**2189-2212.

69. Dobbelaere, S., Vanderleyden, J., Okon, Y. (2003). Plant growth-promoting effects of diazotrophs in the rhizosphere. CRC Critical Review Plant Science,**22,** 107-149.

70. Domingos, M., Klumpp, A., Rinaldi, M.C.S., Modesto, I.F., Klumpp, G., Delitti, W.B.C. (2003). Combined effect of air and soil pollution by fluoride emission on *Tibouchina pulchra* Cogn, at Cubata, SE Brazil, and their relations with aluminium. Plant Soil,**249,** 297-308.

71. Elloumi, N., Abdallah, F.B., Mezghani, I., Rhouma, A., Boukhris, M. (2005). Effect of fluoride on almond seedlings in culture solution. Fluoride,**38(3),** 193-198.

72. Elrashidi, M.A., Lindsay, W.L. (1986). Chemical equilibria of fluorine in soils: A theoretical development. Soil Science,**141(4),** 274-280.

73. Emamverdian, A., Ding, Y., Mokhberdoran, F., Xie, Y.(2015). Heavy metals stress and some mechanisms of plant defense response. Science World Journal, 18Article ID 756120.

74. Faisal, M., Hasnain, S. (2005). Beneficial role of hydrophytes in removing Cr (VI) from wastewater in association with chromate-reducing bacterial strains *Ochrobactrum intermedium* and *Brevibacterium*. International Journal of Phytoremediation, **7,** 271-277.

75. Felestrino, E.B., Santigo, I.F., Freitos, L.S., Rosa, L.H., Ribeiro, S.P., Moreira, L.M. (2017). Plant growth promoting bacteria associated with *Longsdorffiahypogae*Rhizosphere host biological interfera neglected model of bacterial prospection. Frontiers in Microbiology,

doi10.3389ifmicb.017.00172.

76. Finzgar, N., Lestan,D. (2007). Multi-step leaching of Pb and Zn contaminated soils with EDTA. Chemosphere,**66,** 824-832.
77. FRRDF. (1999). State of art report on the extent of F in drinking water and the resulting endemicity in India. Fluorosis Research and Rural Development Foundation, New Delhi.
78. Fung, K.F., Zhang, Z.Q., Wong, J.W.C., Wong, M.H. (1999). Fluoride contents in tea and soil from tea plantations and the release of fluoride into tea liquor during infusion. Environmental Pollution, **104(2),** 197-205.
79. Gangal, R.K. (2007). Geochemical study of groundwater of Sanganer area of Rajasthan (India) and different method to mitigate fluoride problem. Asian Journal of Water Environment Pollution,**4(2),** 93-97.
80. Garrec, J.P., Plebin, R., Audin, M.(1981). Effects of fluoride on the levels of sugars and acids amines of firneedles *Abies alba*. Mill: Environmental pollution series. Aecology Biological,**26,** 281-295.
81. Gautam, N., Verma, P.K., Verma, S., Tripathi, R.D., Trivedi, P.K., Adhikari, B., Chakrabarty, D.(2012). Genome wide identification of rice class I metallothionein gene: tissue expression patterns and induction in response to heavy metal stress. Functional and Integrative Genomics, **12,**635-647.
82. Geebelena, W., Vangronsvelda, J., Adriano, D.C., Poucke, L.C.V., Clijsters, H. (2002). Effects of Pb-EDTA and EDTA on oxidative stress reactions and mineral uptake in *Phaseolus vulgaris.* Physiologia Plantarum **115,** 377-384.
83. Ghosh, P., Rathinasabapathi, B., Ma, L.Q. (2011). Arsenic-resistant bacteria solubilized arsenic in growth media and increased growth of arsenic hyperaccumulator *Pterris vittata* L. Bioresource Technology,**102,** 8756-8761.
84. Giannakoula, A., Moustakas, M., Syros, T., Yupsanis, T. (2010). Aluminum stress induces up regulation of an efficient antioxidant system in the Al-tolerant maize line but not in the Al sensitive line. Environment Experimental Botany, **67(3),** 487-494.
85. Gill, S.S., Khan, N.A., Tuteja, N. (2011). Differential cadmium stress tolerance in five Indian mustard (*Brassica juncea L.*) cultivars, an evaluation of the role of antioxidant machinery. Plant Signal Behavior,**6,** 293-300.

86. Glick, B.R. (1995). The enhancement of plant growth by free-living bacteria. Canadian Journal of Microbiology,**41,** 109-117.

87. Glick, B.R., Penrose, D.M., Li, J. (1998). A model for the lowering of plant ethylene concentrations by plant growth-promoting bacteria. Journal of Theoretical Biology, **190,** 63-68.

88. Glick, B.R., Patten, C.L., Holguin, G., Penrose, D.M. (1999). Biochemical and genetic mechanisms used by plant growth promoting bacteria. Imperial College Press, 1999.

89. Glick, B.R. (2003). Phytoremediation: synergistic use of plants and bacteria to clean up the environment. Biotechnology Advances,**21,** 383-393.

90. Glick, B.R. (2010). Using soil bacteria to facilitate phytoremediation. Biotechnology Advances, **28,** 367-374.

91. Ghnaya, T., Zaier, H., Baioui, R., Sghaier, S., Lucchini, G., Sacchi, G.A. Abdelly, C. (2013). Implication of organic acids in the long-distance transport and the accumulation of lead in *Sesusvium portulacastrum* and *Brassica juncea*. Chemosphere, **90(4),** 1449-1454.

92. Gleba, D., Borisjuk, N.V., Borisjuk, L.G., Kneer, R., Poulev, A., Skarzhinskaya, M., Dushenkov, S., Logendra, S., Gleba, Y.Y. Raskin, I. (1999). Use of plant roots for phytoremediation and molecular farming. Proceeding National Academy Science United States America,**96,** 5973–5977.

93. Goovaerts, P. (1999). Geostatistics in Soil Science: state of art and perspectives, Geoderma, **89,**145.

94. Government of Rajasthan August (2005). State policy for rural drinking water and sanitation, 1-51.

95. Government of Rajasthan (2010). Rajasthan state Environment policy, including Rajasthan environment mission and climate change agenda for Rajasthan 1-122.

96. Grill,E.,Loeffler,S.,Winnacker,E.L.,Zenk,M.H.(1989).Phytochelatins the heavy metals binding peptides of plants are synthesized from glutathione by a specific g-glutamylcysteine dipeptidyl transpeptidase (phytochelatin synthase). Proceeding National Academy Science United States America, **86,**6838-6842.

97. Greipsson, S. (2011). Phytoremediation, National Education Knowledge, **3(10),** 7.

98. Grybos,M., Billard, P., Desobry-Banon, S., Michot, L.J., Lenain, J.F.,

Mustin,C. (2011). Bio-dissolution of colloidal-size clay minerals entrapped in microporous silica gels. Journal of Colloid International and Sciences,**362,** 317-324.

99. Gu, C.S., Liu, L.Q., Deng, Y.M., Zhu, X.D., Huang, S.Z., Lu, X.Q.(2015). The heterologus expression of the *Iris lacteal* var. chinensis type 2 metallothioneninIIMT2b, gene enhances copper tolerance in *Arabidopsis thaliana*. The Bulletin of Environmental Contamination and Toxicology,**94,**247-253.
100. Gupta, S., Banerjee, S., Mondal, S. (2009). Phytotoxicity of fluoride in the germination of paddy (*Oryza sativa*) and its effect on the physiology and biochemistry of germinated seedlings. Fluoride, **42(2),** 142-6.
101. Gupta, S., Banerjee, S. (2011). Fluoride accumulation in crop and vegetable and dietary intake in Fluoride endemic area of west Bengal, Fluoride. **44 (3),** 153-157.
102. Guo, Q., Wang, Y., Ma, T., Ma, R. (2007). Geochemical process controlling the elevated fluoride concentrations in groundwaters of the Taiyuan Basin, Northern China. Journal of Geochemical Exploration,**93,** 1-12.
103. Guo, X.F., Wei, Z., Wu, Q., Li, C., Qian, T., Zheng, W. (2016). Effect of soil washing with only chelators or combining with ferric chloride on soil heavy metal removal and phytoavailability: Field experiments. Chemosphere,**147,** 412-419.
104. Hanger, B.C. (1979). The movement of calcium in plants. Comm. Soil Sciences Plant Analysis,**10(1-2),** 171-193.
105. Harris, H., Hopkinson, D.A. (1976). Handbook of enzyme electrophoresis in Human Genetics, North-Holland, Amsterdam.
106. Hassinen, V.H., Tervahauta, A.I, Schat, H., Ka, S.O.(2011). Plant-metallothioneins-metal chelators with ROS scavenging activity. Plant Biology, 13225-232.
107. Heverkamp,R.G., Marshall, A.T. (2009). The mechanism of metal nanoparticle formation in plants: limits on accumulation. Journal of Nanoparticle Research, **6,**1453-1463.
108. Heidari, M. (2009). Antioxidant activity and osmolyte concentration of sorghum and wheat genotypes under salinity stress. Asian Journal of Plant Science, **8,** 240-244.
109. Hong, B.D., Joo, R.N., Lee, K.S., Lee, D.S., Rhie, J.H., Min, S.W., Song, S.G., Chung, D.Y.(2016). Fluoride in soil and plant. Korean Journal of Agriculture Sciences, **43,** 522-536.
110. Horner, J.M., Bell, J.N.B.(1995). Effects of fluoride and acidity on early plant growth. Agriculture Ecosystems Environment, **52,** 205-211.
111. Hossain, M.A., Ashrafuzzaman, M., Hossain, A.K.M.Z., Ismail, M.R., Koyama, H., Hara, T. (2014). Role of accumulated calcium in alleviating aluminum injury in wheat plants. Science World Journal, 457187.
112. Huang, J., Bachelard, E.P. (1993). Effects of aluminum on growth and cation uptake in seedlings of *Eucalyptus mannifera* and *Pinus radiata*. Plant Soil, **149(1),** 121-127.

113. Huang, J.W., Chen,J., Berti, W.R., Cunningham., S.D.(1997). Phytoremediation of lead-contaminated soils: role of synthetic chelates in lead phytoextraction. Environmental Science and Technology,**31,** 800-805.

114. Huang, G.Y., and Wang, Y.S. (2010). Expression and characterization analysis of type 2 metallothionein from grey mangrove species (*Avicennia marina*) in responses to metal stress. Aquatic Toxicology, **99,**86-92.

115. Hudak, P.F., Sanmanee, S. (2003). Spatial pattern of nitrate, chloride, sulfate and fluoride concentrations in the woodbine aquifer of north-central Texas. Environment Monitoring Assessment, **82,** 311-320.

116. Hussain, J., Hussain, I., Sharma, K.C. (2010). Fluoride and health hazards: community perception in a fluorotic area of central Rajasthan (India): an arid environment. Environment Monitoring Assessment, **162,** 1-14.

117. Indian Council of Medical Research (ICMR). Manual of standards of quality for drinking water supplies, vol. 44. Special Report Series, 2nd ed.; 1975. New Delhi, India.

118. Jariyal, M., Gupta, V.K., Jindal, V., Mandal, K. (2015). Isolation and evaluation of potent *Pseudomonas* species for bioremediation of phorate in amended soil. Ecotoxicology and Environment Safety, **122,** 24-30.

119. Jha, S.K., Nayak, A.K., Sharma, Y.K. (2011). Site specific toxicological risk from F exposure through ingestion of vegetables and cereal crops in Unnao district, Uttar Pradesh, India. Ecotoxicology and Environment Safety,**74,** 940-946.

120. Jha, S.K. (2012). Geochemical and spatial appraisal of fluoride in the soils of indo- gangetic Plows of India using multivariate Analysis, clean Soil, Air, Water, **40,** 1392-1400.

121. Jia, D.U., Jing-Li, Y., Cheng-Hao, L.I.(2012). Advances in metallothionein studies in forest trees. China Plant Omics Journal, **5,**46-51.

122. John, R., Ahmad, P., Gadgel, K., Sharma, S. (2009). Heavy metal toxicity: effect on plant growth, biochemical parameters and metal accumulation by *Brassica juncea L.* International Journal of Plant Production,**3(3),** 65-76.

123. Joshi, V.K. (2003). India's groundwater is flooded with fluoride. Down to earth (Magzine).

124. Kotecha, P.V., Patel, S.V., Bhalani, K.D., Shah, D., Shah, V.S., Mehta, K.G. (2012). Prevalence of dental fluorosis & dental caries in association with high levels of drinking water fluoride content in a district of Gujarat, India. Indian Journal of Medicine Research,**135(6),** 873-877.

125. Kabata-Pendia, A., Pendis, H. (1984). Elements of group VII. In: Trace elements in soils and plants. CRC, Boca Raton, Florida 473-482.

126. Kasai, M., Sasaki, M., Yamamoto, Y., Matsumoto, H.(1992). Aluminum stress increases Kp efflux and activities of ATP- and PPi dependent Hp pumps of tonoplast-enriched membrane vesicles from barley roots. Plant Cell Physiology,**33(7),**1035-1039.

127. Khalid, S., Shahid, M., Niazi, N.K., Murtaza, B., Bibi, I., Dumat, C. (2017). A

comparison of technologies for remediation of heavy metal contaminated soils. Journal of Geochemical Exploration,**182,** 247-268.

128. Khan, G.A., Bauraine, S., Wege, S., Li, Y., deCarbonnel M., Berthamieu, P. (2014). Coordination between zinc and phosphate homeostasis involves the transcription factor PHR1 the phosphate exporter PH01 and its homologue PHO1; H3 in Arabidopsis. Journal of Experimental Botany, 65871-88410.10931jxb/ert444.

129. Khandare, A.L., Rao, G.S. (2006). Uptake of fluoride, aluminum and molybdenum by some vegetables by irrigation water. Journal of Human Ecology, **10(4),** 283-288.

130. Kimura, S., Tamura., K., Nei, M. (2004). MEGA3: Integrated software for molecular evolutionary genetic analysis and sequence alignment. Brief Bioinformatics,**5(2),**150-163.

131. Kim, Y., Kim, J.Y., Kim, K. (2011). Geochemical characteristics of F in groundwater of Gimcheon, Korea: lithogenic and agricultural origins. Environment Earth Sciences, **63,** 1139-1148.

132. Kim, O.S., Cho, Y.J., Lee, K., Yoon, S.H., Kim, M., NaH, Park, S.C., Jeon, Y.S., Lee, J.H., Yi, H., Won, S., Chun, J. (2012). Introducing EzTaxon-e: a prokaryotic 16S rRNA Gene sequence database with phylotypes that represent uncultured species. International Journal of Systematic and Evolutionary Microbiology,**62,** 716-721.

133. Koblar, A.,Tavear, G., Ponikvar-Svet, M. (2012). Fluoride in teas of different types and forms and the exposure of humans to fluoride with tea and diet. Food Chemistry,**130,** 286-290.

134. Komarek, M., Tlustos, P., Szakova, J., Chrastny, V. and Ettler, V. (2007). The use of maize and poplar in chelant-enhanced phytoextraction of lead from contaminated agricultural soils. Chemosphere, **67,** 640–651.

135. Kuang, P., Meng, Y.M., DongChun, Y.S., Fu, Q.H., Yan, L.C., Ma, L. (2016). Effects of nutrition optimization strategy on rhamnolipid production in a *Pseudomonas aeruginosa* strain DN1 for bioremediation of crude oil. Biocatalysis and Agricultural Biotechnology, **6,** 144-155.

136. Kumar, S.P., Prince, W. S. P. M., Sivakumar, S. and Subbhurram, C.V. (2005). *Prosopis juliflora*-A green solution to decontaminate heavy metal (Cu and Cd) contaminated soils. Chemosphere,**60,** 1493-1496.

137. Kumar, K. A.,Varaprasad, P., and Rao, A.V. B. (2009). Effect of fluoride on catalase, guiacol peroxidase and ascorbate oxidase activities in two varieties of mulberry leaves (*Morus alba L.*). Research Journal of Earth Science, **1,** 69-73.

138. Kumar, T., Dhakaand, K.T., Singharya, S.P.(2013). Effect of fluoride toxicity on biochemical parameters (chlorophyll, nitrogen, protein and phosphorus) of wheat (*Triticum aestivum* L.). International Journal of Forestry Crop Improvement,**4,** 80-83.

139. Kumari, S., and Khan, S. (2017). Defluoridation technology for drinking water and tea by green synthesized Fe_3O_4/Al_2O_3 nanoparticles coated polyurethane

foams for rural communities. Scientific Reports, **7,** 8070.

140. Kundu, M.C., Mandal, B. (2009). Assessment of potential hazards of fluoride contamination in drinking water of an intensively cultivated district in West Bengal, India. Environment Monitoring Assessments,**152,** 97-103.

141. Kreig, M.S. (1984). Screening of free living rhizospheric bacteria their multiple plant growth promoter. Journal of Phytopathology, **163,** 173-181.

142. Landis, W.G., Sofield, R.M., Yu, M. (2011). Introduction to environmental toxicology, molecular substructures to ecological landscape, 4th Ed. CRC Press, Taylor & Francis Group, pp. 255.

143. Lee, J. (1983). Gilbert's disease and fluoride intake. Fluoride,**16,** 139-45.

144. Lee, S. (2014). Artificial induction of cadmium tolerance and its further enhancement via heterologous co-expression of SpHMT1 and AtPCS1 in the yeast cells. Journal of Korean Society Applied Biological Chemistry**57,**307-310.

145. Li, L. (2003). The biochemistry and physiology of metallic fluoride: action, mechanism and implication. Critical Review Oral Biology Medicine, **14(2),** 100-114.

146. Li, W.C., Ye, Z.H., Wong, M.H. (2007). Effects of bacteria on enhanced metal uptake of the Cd/Zn-hyperaccumulating plant, *Sedum alfredii*. Journal of Experimental Botany, **58,** 4173-4182.

147. Lin, D.,Xing, B. (2007). Phytotoxicity of nanoprticles: inhibition of seed germination and root growth. Environmental Pollution, **150,**243-250.

148. Lindberg, S.,Wingstrand, G. (1985). Mechanisms for Cd^{2+} inhibition of $(K^{+}+Mg^{2+})$ ATPase activity and uptake in roots of sugar beet(*Beta vulgaris*) Physiology Plant, **63,**181-6.

149. Linger, P., Mussig, J., Fishcher, H., Kobert, J. (2002). Industrial hemp (*Cannabis Sativa L.*) growing on heavy metal contaminated soil: Fibre quality and phytoremediation Potential. Industrial and Crop Products, **16,** 33-42.

150. Lindsay, W.L., Norvell, W.A. (1978). Development of a DTPA soil test for zinc, iron, manganese, and copper. Soil Science American Society Journal, **42,** 421-428.

151. Liu, K., Luan, S. (2001). Internal aluminum block of plant inward K^+ channels. Plant Cell,**13(6),** 1453-1466.

152. Loganathan, P., Hedley, M.J., Grace, N.D., Lee, J., Cronin, S.J., Bolan, N.S., Zanders, J.M.(2003). Fertilizer contaminants in New Zealand grazed pasture with special reference to cadmium and fluorine: a review. Australian Journal of Soil Research,**41,** 501-532.

153. Lominchar, M.A., Sierra, M.J., Millan, R.(2015). Accumulation of mercury in *Typha domingensis* under field conditions. Chemosphere,**119,** 994-999

154. Loper, J. E., Schroth, M.N. (1986). Influence of bacterial sources of indole-2-acetic acid on root elongation of sugar beet. Journal of Phytopathology,**76,** 386-389.

155. Lorestani, B., Cheraghi, M., Yousefi, N. (2011). Phytoremediation potential of native plants growing on heavy metals contaminated soil of copper mine in Iran. World Academy Science Engineering Technology, **77,** 377-382.

156. Lowry, O.H., Rosenbrough, N.J., Farr, A.L., Randall, R.J. (1951). Protein measurement with the folin phenol reagent. Journal of Biological Chemistry,**193,** 265-275.

157. Luck, H. (1974). *In*: Methods in Enzymatic Analysis 2 nd Ed. Bergmeyer Academic Press New York, 885.

158. Lugtenberg, B., Kamilova, F. (2009). Plant-growth-promoting rhizobacteria. Annual Review Microbiology,**63,** 541-556.

159. Luo, Q., Sun, L.N., Wang, H., Hu, X.M. (2015). Metabolic profiling analysis of root exudates from the Cd hyperaccumulator*Sedum alfredii* under different Cd exposure concentration and times. Analytical Methods, **7(9),** 3793-3800.

160. Ma, Y., Rajkumar, M., Freitas, H. (2009). Improvement of plant growth and nickel uptake by nickel resistant-plant-growth promoting bacteria. Journal of Hazards Materials,**166,** 1154-1161.

161. Ma, Y., Rajkumar, M., Luo, Y., Freitas, H.(2011). Inoculation of endophytic bacteria on host and non-host plants-effects on plant growth and Ni uptake. Journal of Hazard Materials,**195,**230-237.

162. Madhavan, N. (2001). Environmental chemistry of fluoride in groundwater and associated soil in Ajmer District, Rajasthan, India. Ph.D Thesis Jawaharlal Nehru University, New Delhi, 4-5.

163. Madhavan, N., Subramanian, V. (2001). Fluoride concentration in river water of a South Asia. Current Science,**80,** 1312-1319.

164. Maegher, R. (2000). Phytoremediation of toxic elemental and organic pollutants. Current Opinion in Plant Biology,**3,** 153-162.

165. Malde, M.K., Scheidegger, R., Julshamn, K., Bader, H. (2011). Substance flow analysis: A case study of fluoride exposure through food and beverages in young children living in Ethiopia. Environment Health Perspect,**119,** 579-584.

166. McQuaker, N.R., Gurney, M. (1977). Determination of total fluoride in soil and vegetation using an alkali fusion-selective electrode technique. Analytical Chemistry, **49,** 53-56.

167. Maestri, E., Marmiroli, M., Visioli, G., Marmiroli, N. (2010). Metal tolerance and hyperaccumulation: costs and trade-offs between traits and environment. Environmental and Experimental Botany,**68(1),** 1-13.

168. Marcele, G.C., Alexandre, C.B., Ray, C., Ana Rasa, R.B., Mathens, P.F., Amanda, S. A. (2014). Effect of cadmium on the content, accumulation and translocation of nutrients in *Bean* plant cultivated in nutritive solution. Communication and Soil Sciences and Plant Analysis, **45,** 223-235.

169. Marques, A.P.G.C., Pires, C., Moreira, H., Rangel, A.O.S.S., Castro, P.M.L. (2010) Assessment of the plant growth promotion abilities of six bacterial species using *Zea mays* as indicator plant. Soil Biology and Biochemistry,**42,** 1229-1235.

170. Mason, B., Moore, C.B. (1982). *Principle of geochemistry*. 4th rd. New York, Wiley, 386.

171. Mathis, P., Kayser, A. (2001). Plant uptake of heavy metals following glyphosate treatment. In: 702 Int. Soc. for trace element biogeochemistry (Ed.), Proc. 6th Int. Conf. on the 703 Geochemistry of Trace Elements (ICOBTE), Guelph, ON, Canada. 29 July-2 Aug. **2001**. 704 Int. Soc. for Trace Element Res., Vienna. 484.

172. Meagher, R.B., Rugh, C.L., Kandasamy, M.K., Gragson, G., Wang, N.J. (2000). Engineered phytoremediation of mercury pollution in soil and water using bacterial genes. In: Terry N, Banuelos G (ed.) phytoremediation of contaminated soil and water. Lewis, Boca Raton, 203 221.

173. Mella, S., Molina, X., Atalah, E. (1994). Prevalence of dental fluorosis and its relation with fluoride content of public drinking water. Revista MedicaChil.e, **122(11),** 1263-1270.

174. Memon, A.R., Schroder, P. (2009). Implications of metal accumulations mechanisms to phytoremediation. Environmental Science and Pollution Research,**16(2),** 162-175.

175. Meenakshi, R.C., Maeshwari, R.C. (2006). Fluoride in drinking water and its removal. Journal of Hazardous Materials,**137,** 456-463.

176. Messaitfa, A. (2008). Fluoride contents in groundwaters and the main consumed foods (dates and tea) in Southern Algeria region. From the issue entitled "Special Issue: Groundwater flow- selected papers from XXXIII IAH Congress, Zacatecas, Mexico (233-320)". Environmental Geology, **55(2),** 377-383.

177. Mirlean, N., Roisenberg, A. (2007). Fluoride distribution in the environment along the gradient of a phosphate-fertiliser production emission (Southern Brazil). Environment Geochemical Health

journal,**29(3),** 179-87.

178. Miller, G.W., Shupe, J.L., Vedina, O.T. (1999). Accumulation of fluoride in plants exposed to geothermal and industrial water. Published by the International Society for Fluoride Research Editorial office, New Zealand.
179. Molla, A.R., Fazlul-Haq, A.K.M., Kazi, M., Fazlul, H. (2007). Measurement of the conc of fluorine in the soil of different areas of Savar and its effects on environment. BRAC University Journal,**1,** 13-17.
180. Mondal, N.C., Prasad, R.K., Saxena, V.K., Singh, Y., Singh, V.S. (2009). Appraisal of highly fluoride zones in groundwater of Kurmapalli Watershed, Nalgonda District, and Andhra Pradesh (India) Environment Earth Science, **59,** 63-73.
181. Moustakas, M., Ouzounidou, G., Lannoye, R. (1995). Aluminum effects on photosynthesis and elemental uptake in an aluminium-tolerant and non-tolerant wheat cultivar. Journal of Plant Nutrition,**18(4),** 669-683.
182. Moustakas, M., Ouzounidou, G., Lannoye, R. (2008) Aluminum effects on photosynthesis and elemental uptake in an aluminum-tolerant and non-tolerant wheat cultivar. Journal of Plant Nutrition,**18(4),** 669-683.
183. Naik, M.M., Pandey, A., Dubey, S.K. (2012).*Pseudomonas aeruginosa* strain WI-1 from Mandovi estuary possesses metallothione in to alleviate lead toxicity and promotes plant growth. Ecotoxicology and Environment Safety,**79,** 129-133.
184. Nandi, J. (2013). Groundwater highly polluted in most parts of Delhi: Report. The Times of India.
185. Neumuller, O.A. (1981). Rommps Chemie Lexicon, Franck'sche Verlagshandlung, Stuttgart (German). 8th ed. 2.
186. Nielsen, M.A. (2009). The unknown treat, implementation of bone chars filters in the area of MEVIWASU together with determination of the lifetime of the filter system, Technical University of Denmark, Department of Environment Engineering.
187. Nichol, B.E., Oliveira, L.A., Glass, A.D.M., Saiddiqi, M.Y. (1993). The effect of aluminum on the influx of calcium, potassium, ammonium nitrate and phosphate in an aluminum sensitive cultivar of barley (*Hordeum vulgare* L.).Plant Physiology, **101(4),** 1263-1266.
188. Nies, D.H. (2003). Efflux-mediated heavy metal resistance in prokaryotes. FEMS Microbiology Review**27(2-3),** 313-339.

189. Niu, Z.X., Sun, L.N., Sun, T.H., Li, Y.S., Wang, H. (2007). Evaluation of phytoextracting cadmium and lead by sunflower, ricinus, alfalfa and mustard in hydroponic culture. Journal of Environmental Sciences,**19,** 961-967.

190. Olsen, S.R., Sommers, L.E. (1982). Methods of determination of phosphorus in soil. American Society Agronomical Madison, WI, USA, 403-430.

191. Oliva, S.R., Mingorance, M.D., Leidi, E.O. (2012). Tolerance to high Zn in the metallophyte Erica and evalensis Cabezudo & Rivera. Ecotoxicol. **21,** 2012-2021.

192. Omueti, J.A.I., Jones, R.L.(1977). Fluoride adsorption by Illinois soils. Journal of Soil Sciences,**28,** 564-572.

193. Panda, D.(2015). Fluoride toxicity stress: physiological and biochemical consequences on plants. International Journal of Bioresearch Environmental Agriculture Science,**1,**70-84.

194. Pandey,A.C., Shekhar, S., Nathawat, M.S. (2012). Evaluation of fluoride contamination in groundwater sources in Palamu District, Jharkhand, India. Journal of Applied Science, **12(9),** 882-887.

195. Pandey, J. (2005). Fluoride distribution and fluorosis in some rural areas of Udaipur. Journal of International Medical Science Academy,**18,** 133-135.

196. Pant, S., Pant,P., Bhiravamurthy, P.V. (2008). Effect of fluoride on early root and shoot growth of typical crop plants of India. Fluoride Journal,**41(1),** 57-60.

197. Patel, M.J., Patel, J.N., Subramnian, S., Eswaramoorthi, S.,Arul, B., Rajeswara, Rao, N., Sridhar, S.G.D. (2001). Distribution of fluoride in the groundwater of Salem and Namakkal districts, Tamil Nadu. Journal of Applied Geochemistry,**3(2),** 120-132.

198. Paulo J.C.F., Pratas, J., Varunc, M., D'Souza, R., Manoj, S., Paul, C. (2014). Accumulation of uranium by aquatic plants in field conditions: Prospects for phytoremediation. Science of the Total Environment, **470(471),** 993-1002.

199. Purcino, A.A.C., Alves, V.M.C., Parentoni, S.N., Belele, C.L., Loguercio, L.L.(2003). Aluminum effects on nitrogen uptake and nitrogen assimilating enzymes in maize genotypes with contrasting tolerance to aluminum toxicity. Journal of Plant Nutrition, **26(1),** 31-61.

200. Putter, J. (1974). In: Methods of enzymatic analysis 2 nd ed. Bergmeyer academic press New York, 685.
201. Quartacci, M.F., Cosi, E.,Navari-Izzo, F. (2001). Lipids and NADPH-dependent superoxide production in plasma membrane vesicles from roots of wheat grown under copper deficiency and excess, Journal of Experimental Botany,**152,** 67-75.
202. Radosław, S., Ciecko, Z.(2017). Effect of soil contamination with fluorine on the yield and content of nitrogen forms in the biomass of crops. Environment Science and Pollution Research,**24,** 8588-860.
203. Rahmana, M.F., Ghosalb, A., Alama, M.F., Kabira, A.H. (2017). Remediation of cadmium toxicity in field peas (*Pisum sativum L.)* through exogenous silicon. Ecotoxicology, Environment Safety,**135,** 165-172.
204. Rajkumar, M., Ae, N., Prasad, M.N.V., Freitas, H. (2010). Potential of siderophore-producing bacteria for improve heavy metal phytoextraction. Trends Biotechnology,**28(3),** 142-149.
205. Reddy, S., Venugopal, N. (1990). Fluoride induced changes in protein metabolism in the tissues of freshwater crab *Barytelphusa guerini.* Environmental Pollution, **67,** 97-108.
206. Rengel, Z., Elliott, D.C. (1992). Mechanism of aluminum inhibition of Net$^{45}Ca^{2+}$uptake by Amaranths protoplasts. Plant Physiology, **98(2),** 632-638.
207. Ribeiro, M.A.Q., de, Almeida, A.A.F., Mielke, M.S., Gomes, F.P., Pires, M.V., Baligar, V.C. (2013). Aluminum effects on growth, photosynthesis, and mineral nutrition of *Cacao* genotypes. Journal of Plant Nutrition,**36(8),** 1161-1179.
208. Robinson, B., Fernandez, J.E., Madejon, P., Maranon, T., Murillo, J.M., Green,S., Clothier, B. (2003). Phytoextraction: An assessment of biogeochemical and economic viability. Plant Soil, **249,** 117-125.
209. Rocca, L.N., Andreoli, C., Giacometti, G.M., Rascio, N., Moro, I. (2009). Responses of the Antarctic microalga *Koliella antartica* (Trebouxiophyceae, Chlorophyta) to cadmium contamination, Photosynthesis, **47,** 471-479
210. Roy, A.K., Sharma, A., Talukder, G.(1988). Some aspects of aluminum toxicity in plants. Botanical Review,**54(2),**145-178.
211. Sabal, D., Khan, T.I., Saxena, R. (2006). Effect of fluoride on cluster bean (*Cyamopsis tetragonoloba*) seed germination and seedling

growth. Fluoride,**39,** 228-230.

212. Sabir, M., Waraich, E.A., Hakeem, K.R., Öztürk, M., Ahmad, H.R., Shahid, M. (2015). Phytoremediation, Soil Remediation and Plants Elsevier Inc. http://dx.doi.org/10. 1016/B978-0-12-799937-1.00004-8.
213. Saini, P., Khan, S., Mamta, B. and Sharma, V. (2012). Organ-wise accumulation of fluoride in *Prosopis juliflora* and its potential for phytoremediation of fluoride contaminated soil. Chemosphere,**89,** 633-635.
214. Saini, P. (2013). In field performance and genetic diversity analysis of *Prosopis juliflora* for fluoride hyperaccumulation. 1-108.
215. Saini, P., Khan, S., Mamta, B., Sharma, V. (2013). Mapping of fluoride endemic area and assessment of F-1 accumulation in soil and vegetation. Environment Monitoring Assessments,**185(2),** 2001-2008.
216. Saleh, S.S., Glick, B.R. (2001). Involvement of gas and pros in enhancement of the plant growth promoting capabilities of *Enterobacter cloacae* CAL2 and UW4. Canadian Journal of Microbiology,**47,** 698-705.
217. Salido, A.L., Hasty, K.L., Lim, J.M., Butcher, D.J. (2003). Phytoremediation of arsenic and 750 lead in contaminated soil using Chinese brake ferns (*Pteris vittata*) and Indian mustard 751 (*Brassica juncea*). International Journal of Phytoremediation, **5,** 89-103.
218. Salt, D.E., Blaylock, M., Kumar, N.P., Dushenkov, V., Ensley, B.D., Chet, I., Raskin, I. (1995). Phytoremediation: a novel strategy for the removal of toxic metals from the environment using plants. Biotechnology,**13,** 468-474.
219. Samra, J.S. (2002) Participatory watershed management in India. J. Indian Soc. Soil Science, **50,** 345-351.
220. Samadi, M.T., Zarrabi, M., Sepehr, M.N., Ramhormozi, S.M., Azizian, S., Amrane, A. (2014). Removal of fluoride ions by ion exchange resin kinetic and equilibrium studies. Environmental Engineering Management Journal,**13(1),** 205-214.
221. Sarkar, D.(2011). Geo-information for appraisal and management of land resources towards optimizing agricultural productivity in the country-I Issues and Strategies. Journal of Indian Society Soil Science,**59,** 35-48.
222. Saxena, S., Rani, A.(2012). Fluoride ion leaching kinetics for alkaline soils of Indian origin. Journal of Science Research Replica,**1,** 29-40.
223. Sehn, P. (2008). Fluoride removal with extra low energy reverse osmosis membranes three years of large scale field experience in Finland. Desalination,**1(3),** 73-84.

224. Schollenberger, C.J., Simon, R.H. (1945). Determination of exchange capacity and exchangeable bases in soil-ammonium acetate method. Soil Science,**59,** 13-24.

225. Sharma, D.C., Pant, R.C. (1994). Chromium uptake its effects on certain plant nutrients in maize (*Zea mays* L.cv ganga 5). Journal of Environment Science health Part,**29,** 941-948.

226. Sharma, R., Bhardwaj, R., Handa, N., Gautam, V., Kohli, S.K., Bali, S., Kaur, P., Thukral, A.K., Arora, S., Ohri, P., Vig, A.P.(2016). Responses of phytochelatins and metallothioneins in alleviation of heavy metal stress in plants:An overview. Plant metal interaction (edited book by Parvaiz Ahmed). 1-651.

227. Shaji, E., Bindu, J.V., Thambi, D.S. (2007). High fluoride in groundwater of Palghat District, Kerala. Current Science, **92,** 240-245.

228. Shen, Z.G., Li, X.D., Wang, C.C., Chen, H.M., Chua, H. (2002).Lead phytoextraction from contaminated soil with high-biomass plant species. Journal of Environment Quality,**31(759),** 1893-1900.

229. Shetty, K.G., Banks, M.K., Hetrick, B.A.D., Schwab, A.P.(1994a). Biological characterization of a southeast Kansas mining site. Water Air Soil Pollution,**78,** 169-177.

230. Shetty, K.G., Hetrick, B.A.D., Figge, D.A.H., Schwab, A.P. (1994b). Effects of mycorrhizae and other soil microbes on revegetation of heavy metal contaminated mine spoil. Environment Pollution,**86,** 181-188.

231. Shortt, H.E., Pandit, C.G., Raghvachari, T.N.S. (1937). Endemic fluorosis in Nellore district of South India. Indian Medical Gazette,**72,** 396-400.

232. Singh, O.V., Labana, S., Pandey, G., Budhiraja, R., and Jain, R. K.(2003). Phytoremediation: An overview of metallic ion decontamination from soil. Applied Microbiology Biotechnology, **61,** 405-412.

233. Singh, R.K., Singh, R.K. (2008). Study of the quality of groundwaters of some villages at Saharsa District, Bihar (India). International Journal Chemical Sciences,**6(3),** 1202-1206.

234. Singh, A.K., Singh, R.S., Shyampura, R. (2008). Soil and water resources of changeri watershed- Udaipur, Rajasthan. Journal of Indian Society Soil Sciences, **56,**106-108.

235. Singh, R.K., Anandhan, S., Singh, S., Patade, V.Y., Ahmed, Z.,Pande, V.(2011). Metallothione-like gene from *Cicer microphyllum* is regulated by multiple abiotic stresses. Protoplasma,**248,**839-847.

236. Singh, U., Arora, N.K., Sachan, P. (2017). Simultaneous biodegradation of phenol and cyanide present in coke-oven effluent using immobilizedPseudomonasputida and Pseudomonas stutzeriogy,http://dx.doi.org/10.1016/j.bjm.2016.12.013.

237. Singh, G., Kumari, B., Sinam, G.G., Kriti, Kumar, N., Mallick, S.(2018). Fluoride distribution and contamination in the water, soil and plants continuum and its remedial technologies, an Indian perspective: a review. Environmental

Pollution,**239,** 95-108.

238. Socha,A.L., Guerinot, M.L. (2014). Mn-euvering manganese: the role of transporter gene family members in manganese uptake and mobilization in plants. Fron. Plant Science, 5.

239. Sood,Anil, Setia, R.K., Bansal, R.L., Sharma, P.K. and Nayyar, V.K.(2004). Spatial distribution of micronutrients in soils of Amritsar district using frontier technologies. Proceedings of 7th Punjab Science Congress (Abstract vol.) held at Guru Nanak Dev University, Amritsar. Feb. 7-9.

240. Spaepen, S., Vanderleyden, J. (2010). Auxin and plant-microbe interactions. Cold Spring Harb Perspect Biol. doi:10.1101/cshperspect.a001438.

241. Shri, M., Dave, R., Diwedi, S., Shukla, D., Kesari, R., Tripathi, R.D., Trivedi,P.K., Chakrabarty, D.(2014). Heterologous expression of *Ceratophyllumdemersum* phytochelatin synthase, CdPCS1, in rice leads to lower arsenic accumulation in grain. Scientific Reports, 4-5784. http://dx.doi.org/10.1038/srep05784.

242. Subbiah, B.V., Bajaj, J.C. (1962). A soil test procedure for assessment of available nitrogen in rice soils. Current Science, 31-196.

243. Sunitha, M.S.L., Prashant, S., Kumar, S.A., Rao, S., Narasu, M., Kishor, P.B.K.(2013). Cellular and molecular mechanisms of heavy metal tolerance in plants: a brief overvie of transgenic plants over-expressing phytochelatins synthase and metallothionein genes. Plant Cell Biotechnology Molecular Biology, **13,**33-48.

244. Sreedevi, P.D., Ahmed, S., Made, B., Ledoux, E., Gandolfi, J.M. (2006). Association of hydrogeological factors in temporal variations of fluoride concentration in a crystalline aquifer in India. Environment Geology,**50,** 1-11.

245. Stampoulis,D., Sinha, S.K., White, J.C.(2009). Assay-dependent phytotoxicity of nanoparticles to plants. Environment Science and Technology,**24,**9473-9479.

246. Stevens, D.P., Mc Laughlin, M.J., Alston, A.M. (1997). Phytotoxicity of aluminium fluoride complexes and their uptake from solution culture by *Avena Sativa* and *Lycopersicon Esculentum*.Plant Soil,**192,** 81-93.

247. Tamura, K., Peterson, D., Peterson, N., Stecher, G., Nei, M., Kumar, S. (2011). MEGA5: molecular evolutionary genetics analysis using maximum likelihood, evolutionary distance, and maximum parsimony methods. Molecular Biology Evolution, **28,** 2731-2739.

248. Tandy, S., Bossart, K., Mueller, R., Ritschel, J., Hauser, L., Schulin, R., Nowack, B. (2004). 768 Extraction of heavy metals from soils using biodegradable chelating agents. Environment Science and Technology,**38,** 937-944.

249. Tangahu, B.V., Sheikh Abdullah, S.R., Basri, H., Idris, M., Anuar, N., Mukhlisin, M. (2011). A review on heavy metals (As, Pb, and Hg) uptake by plants through phytoremediation. International Journal of Chemical Engineering,**31,** 939161.

250. Teotia, S.P., Teotia, M. (1994). Endemic fluorosis in India: A challenging national health problem. Journal of Association Physics India,**32**, 347-352.

251. Tondon, H.L.S. (2002). Method of analysis of soils, plants, waters, fertilizers

and organic manures. Fertiliser development and consultation organisation, New Delhi. India. 204.

252. Tsipa, A., Koutinas, M., Vernardis, S., Mantalaris, A. (2017). The impact of succinate trace on pWW0 and ortho-cleavage pathway transcription inPseudomonasputida mt-2 during toluene biodegradation. Bioresource Technology,**234,** 397-405.

253. Ullah,N., Waqar, A., Munir, H. Z., Zaffar, M., and Weijun, Z. (2014). Enhancing the lead phytostabilization in wetland plant *Juncus effuses L.* through somaclonal manipulation and EDTA enrichment. Arabian Journal of Chemistry, 1878-5352.

254. Usha, B., Prashanth, S.R., Parida, A.(2007). Differential expression of two metallothionein encoding genes during heavy metal stress in the mangrove species, *Avicennia marina* (Forsk.) Vierh. Current Science, **93,**1215-1219.

255. Usha, B., Venkataraman, G., Parida, A., 2009. Heavy metal and abiotic stress inducible metallothionein isoforms from *Prosopis juliflora* (SW) D.C. show differences in binding to heavy metals invitro. Molecular Genetics Genomics, 281, 99-108.

256. Veselov, D., Kudoyarova, G., Symonyan, M., Veselov, S. (2003). Effect of cadmium onion uptake, transpiration and cytokinin content in wheat seedlings. Bulgarian Journal of Plant Physiology,353-359

257. Vestola, E.A., Kuusenaho, M.K., Närhi, H.M., Tuovinen, O.H., Puhakka, J.A., Plumb, J.J., Kaksonen, A.H. (2010). Acid bioleaching of solid waste materials from copper, steel and recycling industries. Hydrometallurgy,**103,** 74-79.

258. Viehweger, K. (2014). How plant cope with heavy metals. Botanical Studies,**55,**35.

259. Vikas, C., Kushwaha, R. K., Pandit, M.K. (2009). Hydrochemical status of groundwater in District Ajmer (NW India) with reference to fluoride distribution. Journal of Geological Society of India,**73,** 773-784.

260. Vikas, C., Kushwaha, R., Ahmad, W., Prasannakumar, V. and Reghunath, R. (2013). Genesis and geochemistry of high fluoride bearing groundwater from a semi-arid terrain of NW India. Environment Earth Science, **68(1),** 289-305.

261. Vladimir, P., Beskoski, A., Cvijovic, G.G., Milic, J., Ilic, M., Miletic, S., Solevic,T., Miroslav M.V. (2011). Ex situ bioremediation of a soil contaminated by Mazut (heavy residual fuel oil)- A field experiment. Chemosphere,**83,** 34-40.

262. Walkley, A., Black, C.A. (1934). An examination of different methods for determining soil organic matter and a proposed modification of the chromic acid titration method. Soil Science, **37,** 29-38.

263. Wang, Y.S., Yang, Z.M. (2005). Nitric oxide reduces aluminum toxicity by preventing oxidative stress in the roots of *Cassia toraL.* Plant Cell Physiology,**46(12),** 1915-1923.

264. Wang, Q., Xiong, D., Zhao, P., Yu, X., Tu, B., Wang, G. (2011). Effect of applying an arsenic-resistant and plant-growth promoting rhizobacterium to enhance soil arsenic phytoremediation by *Populus deltoides* LH05- 17. Journal of Applied Microbiology,**111,** 1065-1074.

265. Wang, B., Wang,Q., Liu, W., Liu, X., Hou, J., Teng, Y., Luo, Y., Christie, P. (2017). Biosurfactant-producing microorganism Pseudomonas sp. SB assists the phytoremediation of DDT-contaminated soil by two grass species. Chemosphere, **182,** 137-142.

266. Whiting, S.N., de Souza, M.P., Terry, N. (2001). Rhizosphere bacteria mobilize Zn for hyperaccumulation by *Thlaspicaerulescens*. Environment Science Technology,**35,** 3144-3150.

267. Willscher, S., Mirgorodsky,D., Jablonski, L., Ollivier, D., Merten, D., Buchel, G., Wittig, J., Werner, P. (2013). Field scale phytoremediation experiments on a heavy metal and uranium contaminated site, and further utilization of the plant residues. Hydrometallurgy, 46-53.

268. Wintzingerode, F.V., Gobel, U.B. and Stackebrandt, E. (1997). Determination of microbial diversity in environmental samples: pitfalls of PCR-based rRNA analysis. FEMS Microbiology Review, **21,** 213-229.

269. WHO(2002). Fluorides. Geneva, Switzerland: World Health Organization. Environmental Health Criteria Number. 227. Accessed in http://www.inchem.org/pages/ehc.html.

270. WHO (1963). World Health Organization International Standards for DrinkingWater. 2nd ed. Geneva.

271. Xia, B., Wang, L., Nie, L., Zhou, Q., Huang, X. (2017). A pathway of bisphenol a affecting mineral element contents in plant roots at different growth stages. Ecotoxicology and Environmental Safety,**135,** 115-122.

272. Yadav, A.K., Khan, P. (2010). Fluoride and Flurosis status in groundwater of Todaraisingh area of District Tonk Rajasthan, India: A case study. Internatational Journal of Chemical Environment Pharmaceutical,**1(1),** 6-11.

273. Yadav, S., Khan, T.I., Gupta, S., Gupta, A.B., Yadava, R.N. (1999). Fluorosis in India with special reference to Rajasthan. In: Proceedings of the international conference on water, environment, ecology, socioeconomics and health engineering (WEESHE), Seoul National University, 3-10.

274. Yang, Y., Han, X., Liang, Y., Ghosh, A., Tang, M. (2015). The Combined Effects of Arbuscular Mycorrhizal Fungi (AMF) and Lead (Pb) Stress on Pb Accumulation, Plant Growth Parameters, Photosynthesis, and Antioxidant Enzymes in *Robinia pseudoacacia* L. PLOS ONE DOI:10.1371/journal.pone.0145726.

275. Yao, Z., Li, J., Xie, H., Yu, C. (2012). Review on remediation technologies of soil contaminated by heavy metals. Procedia Environment Science, **16,** 722-729.

276. Yuan, J., Chen, D., Ren, Y., Zhang, X., Zhao, J.(2008). Characteristic and expression analysis of a metallothionein gene, OsMT2b, down regulated by cytokinin suggests functions in root development and seed embryo germination of rice. Plant Physiology,**146,**1637-1650.

277. Zaidi, S., Usmani, S., Singh, B.R., Musarrat, J. (2006). Significance of *Bacillus subtilis* strain SJ-101 as a bioinoculant for concurrent plant growth promotion and nickel accumulation in *Brassica juncea*. Chemosphere,**64,** 991-997.

278. Zhao, F. J., Lombi, E., Mc Grath, S. P. (2003). Assessing the potential for zinc and cadmium phytoremediation with the hyperaccumulator *Thlaspi caerulescens*. Plant Soil,**249,** 37-43.

279. Zhuang, X., Chen, J., Shim, H., Bai, Z. (2007). New advances in plant growth promoting rhizobacteria for bioremediation. Journal of Environment International,**33,** 406-413.

280. Zsoldos, F., Vashegyi, A., Bona, L., Pecsvaradi, A., Szegletes, Z. (2000). Growth of and potassium transport in winter wheat and durum wheat as affected by various aluminum exposure times. Journal of Plant Nutrition,**23(7),** 913-926.

INDEX

R

S

T

W